'Since 1947, aliens have poured from the abyss that lies between ourselves and the world.'
So begins Bryan Appleyard's dazzling survey of one of the most pervasive yet under-reported phenomena of modern times: aliens – what they are, why they are here, and what they say about us.

Before science and technology took hold of our lives we called them different things: angels, demons, fairies. But in a world in which man has discovered his cosmic insignificance, our need for aliens has reached the stars – and our search for them has drawn our gaze upwards rather than to the bottom of the garden. Since 1947 (Year Zero in the modern history of alienology thanks to a flurry of activity from the first sighting of a 'flying saucer' to the infamous Roswell Incident), there has been a deluge of sightings, abductions, cover-ups, conspiracies and all their associated sub-plots from cattle mutilation to anal probes. Science fiction both on the page and on the screen is in thrall to our perception of and fascination with the Little Green Men – even if, in fact, most of them are Grey.

How did we get here? And what does our obsession with all things alien, whether extraterrestrial or manufactured, say about us in a post-religious world? Venturing fearlessly into the realms of myth, paranoia and the just plain weird, Bryan Appleyard attempts to make sense of the alien phenomenon by exploring not just the close encounters but also the ways in which we have rationalised or celebrated them. Covering everything from the joyful anthropocentrism of *Star Trek* to the bloody-minded nihilism of Stanislaw Le[m] both a brilliantly entertaining cu and an intellectual *tour de forc*

Also by Bryan Appleyard

Understanding the Present:
Science and the Soul of Modern Man

Brave New Worlds: Genetics and the Human Experience

The First Church of the New Millennium

The Pleasures of Peace:
Art and Imagination in Post-War Britain

Richard Rogers: A Biography

Culture Club: Crisis in the Arts

ALIENS

Why They Are Here

BRYAN APPLEYARD

Scribner

First published in Great Britain by Scribner, 2005
An imprint of Simon & Schuster UK Ltd
A Viacom company

Grateful acknowledgement is made to Faber and Faber Ltd for
permission to reproduce lines from 'The Snowman' from *The
Collected Poems of Wallace Stevens*, and to Carcanet Press for
permission to reproduce lines from *Self Portrait in a Convex Mirror*
by John Ashbery.

1 3 5 7 9 10 8 6 4 2

Simon & Schuster UK Ltd
Africa House
64–78 Kingsway
London WC2B 6AH

www.simonsays.co.uk

Simon & Schuster Australia
Sydney

A CIP catalogue record for this book is available from the British
Library

ISBN 0-7432-5685-9
EAN 9780743256858

Typeset by SX Composing DTP, Rayleigh, Essex
Printed and bound in Great Britain by
Mackays of Chatham Plc, Chatham, Kent

For Nigel Andrew

Contents

Acknowledgements

I would like to thank the following for granting me interviews: Graham Birdsall, Georgina Bruni, Michael Buhler, Paul Davies, Chris French, Budd Hopkins, Mike Jay, John Mack, David Oakley, Nick Pope, John Whitmore. I would also like to thank John Gray, Lewis Wolpert and Michael Zimmerman for conversations or assistance. Steven Spielberg unknowingly helped me in the course of an interview given in another context, as did Steve Grand and Stephen Hawking. My daughter, Charlotte Appleyard, helped me with research. David Miller, my agent, did not consider me mad and neither did Andrew Gordon, my editor. My wife, Christena, however, did; she, nevertheless, provided essential support.

Little we see in Nature that is ours

William Wordsworth

The Monster

Since 1947, aliens have poured from the abyss that lies between ourselves and the world. We know them well. They are the grey aliens with slanting, black eyes and vestigial nostrils whose corpses are kept at Area 51, a secret military base in the Nevada desert. They are the tall 'Nordics' who gently oversee complex surgical procedures on human abductees. They are the Borg, the Klingons and the Vulcans, the Pleiadians and the Nine. They are Orthon from Venus, Llysanorh from Mars and Roy Batty in the pouring rain, dying in the arms of Rick Deckard. They are Hal the rogue computer, Klaatu the angry visitor and Gort the avenging robot. There are also the humans – John Mack, Budd Hopkins, Betty and Barney Hill, Whitley Strieber, Kenneth Arnold – who have met or found themselves involved with the aliens. This book is about these people and the real, dreamed, hallucinated or imagined creatures who pour from the abyss to remind us that we can find no home in the world that made us.

These aliens are modern because this is, primarily, the story of the alien invasion that began in the Cascade Mountains in 1947 and which has continued, unabated, ever since. But aliens themselves are not modern. In different guises they have always haunted, tortured, teased, helped, advised or terrorized mankind. Before science and technology took hold of our lives, they were angels, demons, goblins, the Sidhe, fairies, the Hag or any of the billions of spirits that live, half-seen, in the corners of the human world. But, also, they were and still are the inhabitants of the studies and observatories of the philosophers and scientists, the humans who have always speculated about the possibility of extraterrestrial life. They are the life forms that teem among the stars as demanded by Epicurus' Principle of Plenitude. They are the lunarians whose strange way of life was described in detail by the astronomer Johannes Kepler. And they are the Martians that dug the vast canal system spotted by Giovanni Schiaparelli in 1877.

Since 1947, however, the aliens have emerged from the past and from the corners of the world, from the observatories and the studies to stand before us in plain view. Extraterrestrials are now a routine aspect of our culture. Area 51, the place where Fox Mulder's truth is held prisoner in *The X-Files*, appears on TV and in films as the secret site where the corpses and debris of the flying saucer crash at Roswell, New Mexico, are kept. In the United States millions have been abducted by aliens. Millions more have been buzzed by Unidentified Flying Objects. Aliens are the stars of some of the biggest grossing films of all time – *E.T.*, *Alien*, *Star Wars*, *Independence Day* – and of some of the most successful television series – *Star Trek*, *The X-Files*. You can buy inflatable 'Greys' in toy shops, attend UFO conferences around the world or take part in seances to discuss our fate with pan-dimensional beings. Only sex web sites are hit more often than alien ones.

There is also a new academic interest in aliens. NASA now has an astrobiology institute studying the nature of life, not just on Earth but in the cosmos. Steven J. Dick, NASA historian, claims that 'the assertion of extraterrestrial intelligence may be considered a metaphysical completion of the scientific revolution'. Historian Michael J. Crowe of the University of Notre Dame points out that 'even if no extraterrestrials exist, their influence on terrestrials has been immense'. The British philosopher David Lamb has noted the way the search for aliens 'is driven by metaphysical beliefs and a deep psychological desire for companionship'.

But the aliens that surround us are not just extraterrestrials, non-human intelligence may also be possessed by earthly machines. Thinking robots or computers may still lie in the future but machine intelligence is already with us. Computers are so painstakingly adapted to follows our needs and whims that we find ourselves forming virtual relationships with silicon, wire and plastic. Or, out on the net, we chat to the personalities constructed by others, sometimes by machine others. Software, non-humanly, sends us emails. Robot operators offer us 'options' on the telephone. Machines give us money from holes in walls. Daily we interact with a non-human realm that seems to be on the brink of self-consciousness. Indeed, we survive best in the contemporary world by assuming that already the machines are thinking. Bruce Mazlish of Massachusetts Institute of Technology says that our acceptance of a union with our machines will be the 'fourth discontinuity', the latest insult to mankind's groundless sense of self-importance following Copernicus, Darwin and Freud.

And yet, though they are all around us, though non-human intelligence presses in on us as never before, though the history of a plurality of worlds has taken on a new academic seriousness, an interest in aliens is still not quite socially respectable.

'You've really taken on a monster here. Careers are really derailed by an interest in this subject,' I was told by John Mack, professor of psychiatry at Harvard. He speaks from experience. His own interest in alien abductions led to an inquisition by the Harvard University authorities. He survived, though with many scars.

Mention aliens and people will at once think you are either mad or stupid, deluded or naïve. Extraterrestrial visitation is the faith of rural hicks in the American deep South or of lonely English teenage nerds who can't get girlfriends. It is not something serious people with a place in the world should waste their time studying.

A distinction is quietly being made here. People who think there may well be intelligent life elsewhere in the universe are not mad. People who have made contact with aliens, been abducted or who have seen flying saucers probably are. But I'm not mad and I have seen an alien spaceship in broad daylight in Norfolk. It was on the ground and I saw it as a glowing oval behind some trees. It was about 300 yards away. The catch was that I was in a deep hypnotic trance in London at the time. I saw no more than that glow because David Oakley, who had hypnotized me, immediately brought me back from Norfolk to his office. He didn't want to take me any further. He thought any more detailed experience would leave me with a feeling of anxiety about a place I loved. Oakley would say that what I saw was a product of the hypnosis. Others would say it was a buried memory and that I have, in fact, been abducted by aliens who had tried to ensure that I would never remember what happened. The hypnosis retrieved that memory. Who is right?

For the purposes of this book, I would ask you to suspend your impulse to answer that question. Both derisive scepticism and abject faith will obstruct understanding. If you scorn these accounts, you are likely to see them as mere symptoms of individual pathologies and miss the insights they offer. If you accept them uncritically, you will never grasp the complexities involved. Shall we say, for the moment, that the aliens are real? Or, more precisely, that they are *real*.

The italics are simply intended to mark the fact that *something* is going on here. In ordinary speech, the statement 'aliens are real' would be taken to mean that the UFOs, the Greys and the Nordics are as objectively present in the world as my body or this book. That may, indeed, be so. But, even if you do not think that is the case, you can hardly deny, first, that many millions of people do believe they are objectively present and, secondly, that aliens have a genuine and material impact on our lives. Saying the aliens are *real*, therefore, is just a way of encompassing all their possible manifestations and of directing attention to my real concern, which is what the aliens tell us about the human condition.

And so this book is about *real* aliens, who they are and what they want. By aliens I mean intelligent, non-human beings. Such a being may be a machine or an organism. If it is a machine, then it may be a computer or robot made on Earth or some other planet. But, if it is an organism, then it must be from elsewhere. In other words, terrestrial animals do not qualify as aliens for my purposes. Certainly their minds are alien to us and, equally certainly, large-brained, sociable and apparently communicative creatures like dolphins or chimpanzees are possessed of a high degree of intelligence. But they have not yet astounded us with new insights into their own or our condition. The whole purpose of my aliens is to astound us with such insights. They have a problem, however. *Real* as they are, they may not physically exist.

There are, as yet, no intelligent machines. For fifty years scientists have been claiming that terrestrial machines will soon think as well as or better than humans. But computers remain strikingly dumb. Processing speed and power may well exceed that of the human brain within the next twenty years, but, since we still have no idea what thinking is or how it works, these super-fast, super-powerful computers will remain as dumb as ever. Unless, that is, it is not necessary for *us* to know what thinking is. Perhaps these machines will somehow boot *themselves* into self-consciousness. This is what the US defence computer system Skynet does with apocalyptic consequences in the *Terminator* films. Such an event is wildly improbable of course, but, on the other hand, so is human self-consciousness.

The story of Skynet implies that we will not design machine consciousness, it will design itself. It will be an accidental by-product of our ingenuity. Maybe it is already happening. Maybe the vast, virtual 'mind' of the internet is the real world beginning of such a self-booting procedure. Out of the trillions of global connections that mimic the synaptic links in the human brain, a mind will form. It is certainly true that the internet already does things that appear intelligent. But, for the moment, it seems likely that this appearance is solely the work of human minds.

Or, perhaps, we shall turn ourselves into thinking machines. Technological intrusions into the human body may have already turned us into cyborgs: part organism, part machine. From false teeth to pacemakers, we have grown accustomed to such intrusions. There will be many more, raising the question: at what point do we become different kinds of beings? When we do, we will become post-human cyborgs and, therefore, alien. These minds, these imaginations, these souls will be different from those of fully biological beings. They will not be embodied in quite the same way. The alien we have so long sought in the heavens will be returning our gaze from the mirror.

Either of these things may, indeed, be happening. Somewhere a computer may be planning its first move or, after massive surgery, a man has found himself thinking unprecedented thoughts. But we do not know.

Meanwhile, millions of people say that extraterrestrial, intelligent beings are visiting – indeed, have always visited – Earth. We all know what they usually look like, they are the Greys with with large heads and huge black eyes. We also know what alien ships look like – saucer-shaped discs, enormous black triangles or, for me at least, glowing ovals.

We certainly know what these aliens do. They cruise our airspace, scanning our planet. They show particular interest in our military activity and our reproductive system. They also periodically abduct human beings for the purposes of study or for interbreeding with their own species. Some of these visitors mean well, some do not. Some appear to have entered into a secret pact with terrestrial powers, providing advanced technology in exchange for human genes. The deep, official concealment of this pact explains why physical evidence of the existence of extra-terrestrials is unavailable to the masses. It is locked away in government vaults. And alien corpses are kept, frozen, at US air bases.

Many alien visitors, however, are angelic creatures. They are simply here to save us from ourselves and to prepare us for a glorious escape from the muck and mire of human existence and for the loving embrace of the intergalactic community. They may have already performed this service for Marshall Applewhite and his thirty-eight disciples. In March 1997, these members of Heaven's Gate either committed a futile act of group suicide or allowed their non-physical selves to be transported to redemption on board an alien craft that was following the then highly visible comet Hale-Bopp.

Many people, perhaps most, are sceptical of these claims. A few are so sceptical that they devote their lives to proving them wrong, deluded or positive evidence of insanity. If, they say, these aliens are involved in earth-scannings and abductions, why isn't is obvious to us all? The scale of the project seems so vast that it should be on the nightly news. But, plainly, it is not. No alien has chosen to land in Trafalgar Square, on the White House lawn or in the midst of a scientific conference. No indisputably alien artefact is on display in a museum. As the physicist Enrico Fermi asked fifty years ago: where are they? A question that has since been honoured with the title of Fermi's Paradox, though it is not a paradox and the claim of ownership seems excessive since Fermi was not the only man to ask that question.

On the other hand, even the most hardened sceptic will acknowledge that the universe is very big. It is also very strange and growing stranger by the day. At the level of both the very small and the very large, it seems deliberately constructed to defy human reason. Quantum Theory and Relativity both seem to be true but they are also contradictory. Scepticism is not an adequate response to this bizarre construction, unless, that is, it is scepticism of the powers of our own capacity to reason.

In recent years, it has begun to seem less likely that we possess the only self-aware minds in this multi-dimensional paradox. For the first time, we have detected planets orbiting stars other than our sun. On Earth we have discovered 'extremophiles' – creatures that survive in conditions of intense heat or extreme cold, conditions that we previously believed made life impossible. The tenacity of such creatures suggests that life could inhabit extra-terrestrial environments that would once have seemed utterly inhospitable. If, as some believe, whenever conditions are right, life inevitably evolves, then extremophiles suggest the cosmos may be teeming at least with life, if not with intelligent life. After all, life began on Earth extremely quickly – apparently, the ancient rocks suggest, within a couple of hundred million years of its creation.

But the existence, now or in the future, of non-human intelli-gence remains highly contentious. Philosophers and scientists disagree about the viability of a thinking computer. Sceptics and

believers dispute the reality of UFOs, intergalactic communications and alien abductions. Cosmologists and astrobiologists
produce wildly differing figures about the likelihood of life
appearing elsewhere. Some say absolutely yes, others absolutely
no. Most say they don't know.

So are we a staggeringly improbable, freakish, one-off accident
or the universal norm? Are we alone or the unknowing guests at
a cosmic cocktail party? Are we the summit of creation or a
planetful of dirty and violent louts, generally avoided by more
fastidiously evolved beings? Are we the only agglomerations of
matter in the universe that ask such questions?

None of these issues can yet be resolved. Some might never be
resolved – we might, for example, be at that cocktail party but the
guests are too far apart ever to communicate with each other.
And, anyway, it is far from clear how they might be resolved.
Who, for example, should be studying these thousands of reports
of UFO sightings and abductions? Physicists, psychologists or
sociologists? The biologist Edward O. Wilson has cited ufology
as the classic example of a pseudoscience in that it fails to meet
his five necessary criteria of repeatability, economy, mensuration,
heuristics and consilience. But that would seem to be an
admission that there is a whole area in which science is
incompetent. For, as I say, aliens are *real*, something happens to
all these 'experiencers'.

Explanations of the alien phenomenon fall into three
categories: nuts and bolts, third realm and psychosocial. Nuts
and boltists insist that these spacecraft and their occupants are as
real as you and I. They are an aspect of the material world as we
know it, but one that has so far managed to evade the attention
of our mainstream science. Third realmers say they exist, but not
in our reality. They are like angels or demons, neither of the
material world nor of our minds; they inhabit a third realm, a
place we are incapable of fully understanding. Psychosociologists
say they are products of our minds: projections of our anxieties
or aspirations, culturally determined fantasies or hallucinations
or new religions in the process of emerging from the ashes of the
old.

These are not merely explanations of aliens, however, they are world views. They express different attitudes to the peculiar predicament in which we find ourselves – of being alive and consciously so. In defining aliens, we define ourselves. But the definition is, like the aliens, elusive, slippery. What exactly are they? What exactly are we? Perhaps it is the word 'exactly' that is the problem.

My own interpretation, for reasons that will become clear, does not quite fit into any of those three mainstream categories. It is to do with the ambivalent nature of consciousness and our uneasy sense of being both within the world and outside it. This is, in part, a commentary on modernity, but also it is about an eternal aspect of the human predicament that has been massively amplified but not created by modernity.

All of that comes much later. For now, I need simply say that I start from the undoubted *reality* of aliens. They may or may not exist but they are all around us and they are trying to tell us something, possibly about themselves but certainly about us. In order to understand what they are saying it is necessary to abandon the usual barriers between fiction and reality. There are many important connections between the aliens in *Star Trek*, H. G. Wells' Martians, the abductors of Betty and Barney Hill, the cattle mutilators in Montana and the committee of beings known as the Nine who speak through the Florida medium Phyllis V. Schlemmer. These connections form an enormous mirror of ourselves and of our age, through which, like Alice, we can pass and find ourselves in a different world.

This book is about fictional creation and real experience and, on the credibility of the latter, it passes no judgment. It is arranged thematically in order to demonstrate the consistencies and connections that exist across times and genres. For reasons which will become clear, it is primarily about the period since 1947, but it travels back in time to the flying saucers seen by the prophet Ezekiel and to the forgotten civilizations that were seeded from outer space. It is also predominantly about aliens in America. There are two reasons for this: first, though aliens appear all over the world, the primary incidents and documentation are all American and, secondly, television and film have been the forces that have most effectively disseminated and made familiar the great alien tales. Both those industries are dominated by the United States.

There is no one alien narrative, but there is a great body of alien lore. Believers or sceptics, we all partake of this strange knowledge for, in our time, the aliens have stepped out of the shadows and emerged – grey and black-eyed or humming inside a box – as determinants of or commentators on the human condition. John Mack is right. I have taken on a monster. But it is a monster that speaks.

The book is divided into two parts. This is not an absolute division, merely one intended to mark a shift of emphasis. 'Part One: Answers' is primarily about the actual experience of aliens and about the interpretations of experiencers or believers. 'Part Two: Questions' is about the search for aliens, their psychological and metaphysical significance and the imminent arrival of either the intelligent machine or the cyborg. In the conclusion I return to the abyss from which I started.

PART ONE

ANSWERS

1

Kenneth Arnold

On 24 June 1947, Kenneth Arnold, disappointed athlete and successful businessman, saw a squadron of alien spaceships in the Cascade Mountains in Washington state. This is generally regarded as the first authentic Unidentified Flying Object (UFO) sighting of the postwar era. It is also thought by many to be the first use of the word 'saucer' to describe these vehicles. The homely descriptive clarity of the phrase 'flying saucer' and the apparent logic, to the engineering layman at least, of circular ships caught on at once. Alien craft in subsequent Hollywood movies were saucer-shaped.

But, in fact, none of the claims for the exclusiveness or originality of Arnold's sighting is correct. It had been preceded by the postwar appearance in Sweden of 'ghost rockets' – craft that flashed across the sky, finally disappearing into lakes or the sea. Given the proximity to the border, these were at first taken to be Soviet weapons tests, a theory later discounted. In November 1948 a top-secret Swedish report concluded that 'fully technically qualified people have reached the conclusion that these phenomena are obviously the result of a high technical skill which cannot be credited to any known culture on earth'.

There had also been sightings of 'ghost aeroplanes' during the war in Britain. And, in the US, there had been several UFO reports before Arnold's in Montana, Connecticut, Virginia and Wisconsin. Project Sign, the US Air Force's first investigation of UFOs, actually classified Arnold as 'Incident #17'. During the war there had also been the 'foo fighters', glowing spheres that seemed to pursue military aircraft. On one occasion a foo fighter penetrated the body of an American bomber and travelled slowly up and down the interior before leaving through the tail. These had been regularly reported by pilots and, like the ghost rockets, they remain unexplained.

In truth, by the time of Arnold's sighting, extraterrestrial spacecraft seemed to have been arriving for almost a century. In June 1864 in France a newspaper reported that American geologists had discovered a hollow, egg-shaped object containing the mummified body of a three-foot humanoid creature with a trunk projecting from the middle of its forehead. In 1884 in Nebraska cowboys saw what was taken to be an airship from another planet crash to earth. A wave of cigar-shaped UFO sightings began in northern California in 1896 and ended in the Midwest six months later. A curious letter in the *Otago Daily Times* in New Zealand on 29 July 1909 confidently stated that local airship sightings were of 'atomic-powered spaceships' from Mars.

'From about 1880 through 1946,' writes the folklorist Thomas E. Bullard, 'witnesses reported a stream of aerial mysteries – phantom airships, mysterious airplanes, foo fighters and ghost rockets – that reflected technological expectations of the day or looked one step ahead. The phantom airships of 1897 played to human hopes by heralding that man could fly at last, whereas foo fighters as supposed enemy secret weapons during World War II and ghost rockets as possible Soviet missiles at the onset of the cold war fueled human fears of destruction from the sky.'

Bullard, psychosocially, suggests the objects were generated by the climate of the times working on human imaginations. But, whatever the explanation, there had been many strange things in the sky and many people had concluded that they were, indeed, alien craft.

The incidents, however, were isolated oddities until June 1947. It was then that something previously blurred and indistinct began to appear in sharp focus. Discrete incidents began to coalesce into a distinct phenomenon. There was a mood of expectation, even before Arnold. As if in answer to cosmic promptings, just two days before the sighting, Dr Layman Spitzer, an astrophysicist at Yale, had appeared on a radio programme to suggest that Earth might already have been visited by Martians.

'It is possible,' Spitzer told listeners to *Yale Interprets the News*, 'that the Martians may have been civilized for millions of years as compared to our thousands. In such a case, their scientific knowledge, and their ability to control nature would, of course, be enormously greater than ours.'

Spitzer speculated that the Martians might not have been spotted because they had not spent time in a large city and had not been photographed. Anybody who had seen them, he mused presciently, would probably not be believed. The stage was being set for the great postwar alien 'flap'.

As for the saucer simile, that also was not exclusive to Arnold. John Martin, a farmer from Denison, Texas, saw a UFO during a hunting expedition on 2 January 1878 and said it was 'about the size of a large saucer' when it was directly overhead. Martin used 'saucer' to describe the size; its shape, he thought, was that of a balloon.

'Mr Martin is a gentleman of undoubted veracity,' reported the conscientious *Denison Daily News* on 25 January, 'and this strange occurrence, if it was not a balloon, deserves the attention of our scientists.'

In fact, Arnold's use of the word 'saucer' was not quite what it seemed either. As in Martin's report, it was nothing to do with the shape of the objects. He told a local journalist, Bill Bequette, they were flying 'like a saucer if you skipped it across the water'. Artists' impressions based on his descriptions later showed not circular, but curved, boomerang-shaped craft. It was the newspapers that coined the imaginatively gripping phrase 'flying saucer'.

This confusion points to a fundamental and enduring

ambiguity about UFOs. Are they, in any sense at all, real or are they products of our imagination?

Neither Martin nor Arnold had reported a saucer-shaped craft and yet the use of the word saucer had created a powerful, popular image whose ascendancy in the world of ufology would be challenged only by the much later sightings of enormous, triangular ships of the type routinely seen by Agent Fox Mulder in *The X-Files*. For the sceptic, this indicates that the UFOs were human inventions. It was the use of the word 'saucer' that had created saucer-shaped ships in the imaginations of the gullible. For the believer, it is compelling evidence of the objective truth of alien visitations. These ships are consistently said to be 'like' saucers because that is, indeed, the shape of alien craft. In the odd world of alienology, consistency of observation seems to work equally well for both the prosecution and the defence. And, the believer might add, even if the details are wrong, the fact that sober observers see inexplicable things is beyond dispute.

Either way, Arnold was neither unique nor original. In addition, subsequent analysis of his sighting by ufologists has cast doubt on its reliability. In part, this was because this is generally seen as a solo sighting. The pilot of a nearby DC4 saw nothing. There was, apparently, one Fred M. Johnson, a prospector, who endorsed the account. But his name does not seem to occur very often in the Arnold literature.

Furthermore, close analysis of Arnold's report revealed inconsistencies. Arnold said the objects seemed about twenty times as long as they were wide. Some have pointed out that, as he said they were 25 miles away and there are known limitations to the resolving powers of the human eye, this would have made the objects 100 feet thick and 2000 feet long. This is implausible, though not, of course, impossible.

After the sighting, Arnold's capacity for objectivity was also called into question. In an essay entitled 'Resolving Arnold', psychologist Martin Kottmeyer summarizes the problem in a way that reveals the exotic and frequently profound issues raised by the attempts to verify UFO sightings as well as the strange world into which Arnold had stepped as a result of his sighting (see www.reall.org). The essay does not question Arnold's sincerity but does point out the eccentricity of some of his later claims – that the UFOs were space animals with the ability to change their density, that in 1952 he saw two living transparent UFOs that seemed to be aware of him.

'This sounds,' observes Kottmeyer, 'suggestive of a delusion of observation and the possible presence of paranoia . . . It is reasonably probable that this paranoid cast of thought was rooted in a knee injury which thwarted Olympic ambitions and blew apart plans he had to use his athletic talent to forward his college education. It is a common syndrome that has been termed "the athlete's neurosis". Such an incident could inculcate a habit of emotionally generated misinterpretation.'

Kottmeyer also mentions that Arnold's later claims to have had other sightings were seized upon by sceptics as evidence of his unreliability. But, again, this kind of evidence cuts both ways. Later accounts by abductees seem to indicate that being a 'repeater' is, in fact, a clear sign of authenticity. The aliens appear in front of or abduct the same people over and over again.

There are also various sceptical theories about what Arnold actually saw. A flock of American white pelicans – huge birds with wingspans of 110 inches – is one possibility. Keay Davidson, a science writer for the *San Francisco Examiner*, suggested they were meteor fragments. In the Northern Hemisphere there were indeed spectacularly high levels of meteorite sightings in June 1947. Nevertheless, in spite of inconsistencies, eccentricities and the best efforts of the debunkers, Arnold's sighting of the ships still stands as the initiator of the modern era of UFO sightings and extraterrestrial interventions.

'The news flashed around the world with the speed of light waves,' Desmond Leslie wrote, 'and started the commotion we call flying saucers.' The eccentric, Anglo-Irish Leslie was an RAF pilot, author, composer, film-maker, nightclub director, estate manager and spiritualist, but above all he was a ufologist, a 'pioneer of the unusual' in the words of the *Irish Times* obituary in 2001. In 1953 he co-wrote *Flying Saucers Have Landed* with George Adamski.

The post-Arnold commotion was startling. The Roswell incident, involving a crashed flying saucer, happened just two

weeks after the Arnold sighting. At the FBI, director J. Edgar
Hoover was complaining that he knew the military had got hold
of a saucer 'and would not let us have it for cursory examination'.
In the first two weeks of July the Air National Guards of the
Pacific Coast States flew 'saucer patrols'. Oregon had eight P-51
fighters and three A-26 bombers aloft looking for saucers.
Military aircraft pursued and lost one in North Dakota. On 10
July mechanics at Harmon Field, Newfoundland, photographed
a brilliant disc that cut through the clouds. And so on for the rest
of 1947; indeed, for the rest of the postwar era.

Oddly, in retrospect, it was not immediately assumed by
everybody that these discs contained extraterrestrials. A Gallup
poll conducted in August 1947 asked people what they thought
was behind the flying saucer sightings. Some 33 per cent said they
didn't know or gave no answer, 29 per cent said they were the
products of imagination, or optical illusions or mirages, 15 per
cent believed they were US secret weapons, 10 per cent thought
they were hoaxes, 3 per cent thought they were weather
forecasting devices and 1 per cent thought they were Russian
secret weapons. The extraterrestrial hypothesis did not register in
the findings.

Nevertheless, the idea was out there. On 3 July – after Arnold
but before Roswell – Ole J. Sneide wrote to the *San Francisco
Chronicle* to explain that the 'flying disks are oblate spheroid
space ships from the older planets and other solar systems . . .
Their present local headquarters is on the unseen side of our
moon.' Sneide had apprised himself of these details while
teleporting himself about the galaxy.

Arnold had initiated the modern UFO era. But why this case
among so many others? Because, whatever he got up to later,
Arnold, at the time, appeared to be an excellent witness, a
reassuringly normal man. There was also an eerie clarity about
his sighting.

His report to the American military authorities, submitted on
12 July 1947, could scarcely have been more plausible. Aged
thirty-two at the time of the sighting on Tuesday 24 June, he had
been a fine athlete at school and subsequently a successful
businessman. He was also a very experienced pilot – 'To date,' he

wrote in the report, 'I have landed in 823 cow pastures in mountain meadows, and in over a thousand hours a flat tire has been my greatest mishap.'

In January he had bought a new Callair plane, specially designed for 'high altitude take-offs and short rough field use'. It was in this craft that, at 2pm on 24 June, he took off from Chehalis in Washington state to fly to Yakima. His take-off had been delayed for an hour because of a marine transport craft that seemed to have gone down near Mount Rainier. This had not been found by the time Arnold wrote his report, suggesting that an abduction event may also have been involved. During the flight, he made a few passes around Mount Rainier, looking for the lost plane. Finally, at an altitude of 9200 feet, he was heading back towards the mountain. The air was crystal clear. After a few minutes on this course, he saw a bright flash. He could not trace its source until he looked left to the north of the mountain.

'I observed a chain of nine peculiar looking aircraft flying from north to south at approximately 9500 foot elevation and going, seemingly, in a definite direction of about 170 degrees.'

His precision was remarkable and in sharp contrast to the vagaries of most earlier and later UFO reports. At first he assumed they were military jets. They were moving quickly but not, he thought, at a speed that would be beyond the capabilities of the new aircraft then being tested. But then, as they approached Mount Rainier, he saw them outlined against the snow and he noticed they had no tailplanes. The craft maintained a rough formation, but, individually, they would occasionally dip and change course slightly – hence his later description of their flight pattern. In total, he watched them for between two and a half and three minutes. At times, when they caught the sun, they appeared to be completely round.

There was something clear, sunlit and exact about Arnold's encounter. In the bright mountain air of the Cascades, against the snowfields of Mount Rainier, it seemed he could not be mistaken. The few minutes in which he tracked that mysterious squadron were lucid, crystalline. The man himself seemed equally clear – normal, super-normal perhaps, an ordinary man to whom an extraordinary thing had happened. If people like him saw strange things, then there must, indeed, be strange things.

Because of this clarity and exactness, the Arnold sighting represents a turning point. Of course, strange things had been seen in the sky throughout history and, of course, UFOs of various kinds had been seen regularly in the days preceding 24 June 1947. But previous reports of alien craft had been mere eccentricities, patternless mysteries that intrigued and entertained a few connoisseurs of the abnormal. But, with Arnold, all of that changed. June 24th was a day on which something previously unformed, something that was both an anxiety and an aspiration, sharpened into a coherent image in the contemporary imagination. The aliens had arrived. But why had they come? Why now?

Simple. They were angry.

2

Alien Indignation

'We have come to visit you in peace and with goodwill,' says Klaatu. The soldiers cock their pistols. He walks towards the waiting crowd of humans and starts to retrieve something from inside his silver one-piece suit. A soldier, thinking he's going for his ray gun, panics and fires. Klaatu drops to the ground, wounded. But it isn't a ray gun, it is an object with spiky aerials at one end. It is a gift of peace, a communication device for the President of the United States that would allow him to study life on other planets.

Now Gort emerges from the flying saucer. He is a robot, massive, metallic. A strange visor conceals his eyes. He stops, the visor rises and a beam flares out. It destroys the guns and tanks of the soldiers.

Klaatu, who looks like a tall, thin, handsome human, wears a shiny, loose-fitting, one-piece suit. He has travelled for five months and covered 250 million miles to reach Earth, only to be shot when he does so. He may well have been expecting something of the sort. His people have been monitoring our planet for years and have noted our aggressive tendencies. But now they can no longer just watch our antics. Something has changed. Klaatu has an urgent message for our leaders. It must be delivered to all of them at once and so they must be gathered at one place. This requirement turns out to be beyond human diplomacy. The leaders are too quarrelsome to agree on a meeting.

But we do, finally, learn what the message is when, on departure, Klaatu addresses a gathering of scientists from the ramp of his flying saucer. Humans have recently discovered atomic energy and invented rockets – this is 1951 – and the aliens' monitoring had made them all too aware that we are a bellicose species. This would not matter if we merely had our old tanks and guns and used them only to kill each other. They are primitive weapons. But these new technologies are more dangerous; they are a step upwards and outwards. They make us a threat to our galactic neighbours. We must now learn to behave more in keeping with the standards of the cosmic community. If we do not, our planet will be eliminated.

'If you threaten to extend your violence, this earth of yours will be reduced to a burned-out cinder,' warns Klaatu.

Gort, it turns out, is the means whereby other civilizations have controlled their own violent impulses. He is a member of a robot police force that is programmed to disarm any aggressor. The overwhelming power of the robots means that conflict is impossible. Gort is implacable and indestructible. He renders diplomacy futile since he is not open to negotiation. He is both the servant and the master of his makers. A metal Jeeves, he both does their bidding and saves them from themselves.

This film – *The Day the Earth Stood Still* (Robert Wise, 1951) – is based on a short story. But in 'Farewell to the Master' by Harry Bates, published in *Astounding* magazine, there is a twist ending, omitted in the movie. The twist is that Gort is the master, Klaatu the servant. The movie has less complex intentions.

Humanity does not understand anything of Klaatu's message. His sojourn on Earth is violent and eventful. He is subsequently shot yet again, this time fatally. But he is brought back from the dead by a machine in his saucer operated by Gort.

At last, the time comes for Klaatu to leave. He waves to the crowd from the ramp of his saucer and catches the eye of Helen Benson, the girl who has begun to fall in love with him. His shining, alien goodness has exposed her human boyfriend for what he is – a cheap, conscienceless grifter. She, at least, has seen the error of her ways; so, too, must all other human beings. The aliens are better than us.

(Curiously, the central idea in this film, that an encounter with advanced alien technology would invalidate Earth's war machines, was anticipated by one of the twentieth century's most brutal advocates of political violence. In conversation with H. G. Wells, Vladimir Ilyich Lenin said:

'You are right. I understood this myself when I read your novel *The Time Machine*. All human conceptions are on the scale of our planet. They are based on the pretension that the technical potential, although it will develop, will never exceed the "terrestrial limit".'

'If we succeed in establishing interplanetary communications, all our philosophies, moral and social views will have to be revised. In this case the technical potential, become limitless, would impose the end of the role of violence as a means and method of progress.'

The arrival of the aliens would change everything, even Lenin's steely certainties. Aliens relativized our most cherished absolutes. But not entirely. The robot's justice was a form of twentieth-century realpolitik. Had Lenin lived, he would have liked the idea of Gort.)

The Day the Earth Stood Still was the second alien visitor film of the fifties, the first being *The Thing from Another World* (Christian Nyby, 1951). Both films involve flying saucers and both appeared just four years after the Arnold sightings. Aliens and saucers were plainly on the popular agenda. But so too was alien indignation at human behaviour.

In those four years, all hell had broken loose. Barely two weeks after Arnold, an official press release concerning an incident at the Roswell Army Air Field in New Mexico said that 'a disc' had been recovered. The *Roswell Daily Record* of 8 July was able to run one of the most sensational intros in the history of journalism.

'The intelligence office of the 509th Bombardment group at Roswell Army Air Field announced at noon today, that the field has come into possession of a flying saucer.'

The aliens had crash-landed and, according to reports more than thirty years later, their machinery and corpses had been locked away for study by the US military. But the first question was: what were they doing flying around Roswell?

The answer was glaringly obvious. The 509th was the bomber group that dropped the A-bombs on Hiroshima and Nagasaki and was, therefore, the only United States Air Force unit with nuclear experience. The aliens were obviously checking out our most advanced weaponry. Like Klaatu, they were concerned at this extension of our military capabilities. Checking out the military was something the aliens would continue to do, a fact that obliged the military and intelligence communities to pay at least some attention to the mounting number of UFO reports.

'At this time,' wrote H. Marshall Chadwell, a scientific intelligence officer, in a memo to the director of the CIA in December 1952, 'reports of incidents convince us that there is something going on that must have immediate attention . . . Sightings of unexplained objects at great altitudes and travelling at high speeds in the vicinity of major US defense installations are of such nature that they are not attributable to natural phenomena or known types of aerial vehicles.'

Alien interest in military establishments has continued ever since. As Klaatu made clear, it is our weapons that concern them. The Rendlesham Forest incident – 'Britain's Roswell' – happened near two air force bases in Suffolk in 1980. UFOs landed and were seen by air force personnel, in some accounts aliens were actually encountered. And, in 1975, UFOs were seen buzzing silos housing Minuteman intercontinental ballistic missile silos in Montana. These latter visitations were also accompanied by cattle mutilations: mysterious and fatal surgical operations on local livestock.

Meanwhile, back in 1947, the US Air Force reported 122 cases of UFO sightings in that year. This was an absurdly conservative figure. One study of local newspaper reports puts the number at 850, just in the period 1 June to 31 July. Aliens were gleefully, not to say capriciously, buzzing the entire continent.

'They behave,' wrote the psychiatrist Carl Gustav Jung in his book of 1958, *Flying Saucers: A Modern Myth of Things Seen in the Sky*, 'more like groups of tourists unsystematically viewing the countryside, pausing now here for a while and now there, erratically following first one interest and then another, sometimes shooting to enormous altitudes for inexplicable reasons or performing acrobatic evolutions before the noses of exasperated pilots.'

Opinion polls suggested that, within months, 90 per cent of the US population had heard of flying saucers, though most people thought they were explicable in earthly terms. But the media were definitely very keen on the extraterrestrial explanation. First there were the newspapers, then the books – Arnold himself co-wrote *The Coming of the Saucers* with Ray Palmer, managing editor of the SF magazine *Amazing Stories* – and then the films.

By the time Hollywood took notice, the sheer weight of UFO reports clogging papers across the United States had inspired a certain weariness. In the opening sequence of *The Day the Earth Stood Still* a radio announcer insists: 'This is not another flying saucer scare, scientists and military men are already agreed on that.' Clearly there had been so many saucer scares that the public could not be relied upon to recognize the real thing when it happened.

But, in 1947, the real thing seemed to be happening all over the place. The sky was full of saucers and so, finally, on 22 January 1948, unable to take any more of this, the USAF established an investigation, Project Sign, to try to find out what was going on. After all, if mysterious craft were using US airspace with impunity, it was a serious defence matter. The order establishing the project stated that its aim was '. . . to collect, collate, evaluate and distribute to interested government agencies and contractors all information concerning sightings and phenomena in the atmosphere which can be construed to be of concern to the national security'.

Sign was to be the only official investigation of UFO reports in the US to give some credence to the possibility that these things were, indeed, alien ships. Later studies – Project Grudge, established in December 1948, and Project Blue Book, established in March 1952 and continuing until 1969 – were invariably sceptical of anything but earthbound explanations.

With Sign under way, with Arnold, with Roswell and with the 1947 wave of sightings, the great UFO 'flap' that has now lasted more than sixty years had officially begun. And, as *The Day the Earth Stood Still* demonstrates, the first and still the dominant theme of this flap had been defined almost at once. That theme was human guilt, guilt at what we were, guilt at what we had become, guilt at what we were doing. Klaatu was unambiguous: we were the hooligans of the cosmic community. The aliens were here because our technology had crossed the line. Always dangerous to ourselves, we were now a danger to others. The aliens were indignant at our behaviour, appalled by our violence and brutality and, like Klaatu, they were determined to do something about it.

This and the accompanying fear that we would infect other civilizations became an artistic convention. In Ray Bradbury's *The Martian Chronicles*, published in 1951, Spender contemplates the horrific possibility that humans will transfer their vile habits to Mars.

'What could I do? Argue with you? It's simply me against the whole crooked grinding greedy set-up on Earth. They'll be flopping their filthy atom bombs up here, fighting for bases to have wars. Isn't it enough they've ruined one planet, without ruining another? Do they have to foul someone else's manger? The simple-minded windbags. When I got up here I felt I was not only free of their so-called culture, I felt I was free of their ethics and their customs. I'm out of their frame of reference, I thought. All I have to do is kill you off and live my own life.'

Bradbury had captured the climate of guilt and self-loathing that had infected the world. Why? Because, suddenly, we seemed to have gone too far.

The exact timing is crucial. The modern UFO/alien flap began in a year of wonders, mysterious portents and terrible anxieties. Nineteen forty-seven was an astonishing year in which people began to believe anything was possible. It was the start of an astonishing era in which all certainties crumbled into dust.

'Nuclear physics,' wrote Jung, 'has begotten in the layman's head an uncertainty of judgement that far exceeds that of the physicists, and makes things appear possible which but a short while ago would have been declared nonsensical.'

It was, for a start, the first full year in which the possibility of nuclear annihilation had begun to sink into the human imagination. The Hiroshima and Nagasaki bombs had been the initial warning, of course, but then, in summer 1946, came Operation Crossroads. This involved the detonation of two 23-kiloton atom bombs, codenamed Able and Baker, at Bikini Atoll in the Pacific. With the first test at Alamogordo, New Mexico, in 1945 and the two Japanese attacks, this brought to five the total number of nuclear explosions on Earth.

Crossroads was intended to test the effect of nuclear explosions on shipping. Needless to say, the effect was dire. Nevertheless, its location resulted in the naming of a two-piece swimsuit. This close proximity of frivolous sexual provocation and appalling destruction was a poignant summary of the conflicting themes of the age. The bikini was a symbol of the good, consumer life that was on the horizon, bringing washing machines, televisions and vacuum cleaners. The fifties was to be the decade in which the consumer society took off, promising a new paradise of leisure attained and labour saved. But the bikinis and washing machines were the apples in the new Eden. The Bomb was God's judgment. We ate and found ourselves evicted into a blasted, alien landscape of ultimate risk.

As early as 1951, this proximity of destruction and consumption had begun to inspire guilt, an intuitive awareness that the Bomb was the price we had to pay for our sins of excess and that the price was too high. That year Frederick Pohl's short story 'The Midas Plague' appeared in *Galaxy* magazine. Set in the future, it portrays a society in which everybody is legally obliged to consume as much as possible. Robots produce a surplus of goods that have to be disposed of by the assiduously consuming masses. Morey, the hero, solves his consumption problem by closing the circle. He teaches his domestic robots to consume so he doesn't have to. Sadly, he is found out. Consumption, like the Bomb, threatens to end civilization.

But atom bombs were more aesthetically gripping than shops. They were surreal destroyers, drunken conjurers or clowns,

intent on randomly reforming matter. The sand at Alamogordo
was turned into glass by the heat of the Trinity test, a Daliesque
hardening of softness that unnervingly echoes the poignant
conjunction of the bikini and the Bomb. And I remember, when
young, reading an account of the aftermath of Hiroshima. A man
grasped a woman's hands to help her and the skin came off 'like
gloves'. The exactness of the simile replaced the horror with
an eerie elegance and gracefulness as well as a very tasteful
eroticism. The world of the Bomb is a dream world, an alien
world, and now it was our world.

And so, evicted from Eden, we found ourselves in this bleak,
improbable dream-prison, filled with unthinkable horrors, with
skin like gloves and sand like glass. But, wrote Jung, 'humanity
would like to escape from its prison'. How? The affluence and the
apocalypse seemed to be two sides of the same melted coin.

The novelist Stephen King was later to note this eerie
convergence in this postwar/Cold War era:

> We were fertile ground for the seeds of terror, we war
> babies; we had been raised in a strange circus atmosphere
> of paranoia, patriotism and national hubris. We were told
> that we were the greatest nation on earth and that an Iron
> Curtain outlaw who tried to draw down on us in that great
> saloon of international politics would discover who the
> fastest gun in the West was . . . But we were also told exactly
> what to keep in our fallout shelters and how long we would
> have to stay in there after we won the war. We had more to
> eat than any nation in the history of the world, but there
> were traces of strontium-90 in our milk from nuclear
> testing.

These bombs were not just big explosions, they represented a
human assault on nature that was almost comically fundamental.
They produced an invisible killer, radiation, that was emblematic
of the way that nuclear power seemed to represent a tearing of the
visible veil of the world. Radioactive strontium-90 was particularly disturbing, it infected the milk of motherhood with its
death rays. This, surely, had to be some kind of joke.

It was. It was the eternal joke of human overreaching. There were many further portents that we had gone too far. Also in 1947, on 14 October Chuck Yeager flew faster than sound in his Bell X-1. He was said to have 'broken the sound barrier', another veil had been torn. The very phrase 'sound barrier' has something ominously paradoxical about it, like 'light year', 'virtual reality' or 'real time'. Such phrases betoken a new reality that seems determined to offend the old by warping the language into irrational configurations. Should this sound barrier have been broken? Is that what we now did? Break the stuff of nature? One side of the human imagination welcomed such things as great adventures; the other side feared impiety in the face of creation.

That strange year went on. On 17 November Walter Brattain dropped a silicon contraption into a thermos of water to rid it of condensation. Once wet, the device produced a prodigious electronic amplification. Brattain and his colleague, Robert Gibney, stared in disbelief. They had invented the transistor and the information age, the age of quick, electronic immateriality that would supersede the age of heavy, ironclad machines.

We were, though still in our dream-prison, moving upwards into another realm. There was a sense of transition to higher forms of technology. And that, combined with the certainty that there would be new, more terrible wars, fired a passion for novelty and innovation. In October 1945 Project Rand (standing for research and development) was set up under the auspices of the Douglas Aircraft Company. In May 1946 Rand issued its first report. It was entitled 'Preliminary Design of an Experimental World-Circling Spaceship'. Old battlefields were being left behind; space was the exciting new one. Rand was later to make its name by devising 'winnable' strategies for nuclear wars.

Flying saucers were an aspect of a widespread belief in the possibility of radical and exotic technologies. At Wright-Patterson Air Force Base they were working on disc-shaped aircraft and at Cal Tech they were researching the Coanda effect. This had been discovered in 1930. It was based on the way fluid tends to follow the curvature of a surface and could, conceivably, provide propulsion for a saucer-shaped craft. At Wright-Patterson they were also looking into the German flying wings designed by the Horten brothers, machines that anticipated the Stealth bomber and which seemed to be taken straight from the pages of *Astounding* magazine. A different world lay on the far side of the torn veil.

There was one further startling and momentous invention in 1947. The National Security Act was passed in the US Congress. It created the Central Intelligence Agency, the quasi-demonic entity that was to become the focus of all paranoia arising from the increasingly widespread conviction that the aliens had landed and we hadn't been told. Again the words resonate disturbingly, 'central intelligence' suggesting some vast, higher brain had been created.

Perhaps it had. Why not? We had broken through to the other side and now anything was possible and, thanks to the CIA, anything might be hidden from democratic scrutiny. Incredulity – about aliens or technology – was no longer a reasonable position. In the decade that followed the year of wonders, the message was slowly absorbed. In the film *Them!* (Gordon Douglas, 1954), for example, atomic radiation causes the emergence of giant ants in New Mexico, a place that had rapidly become an alien hotspot. The point was that this weird, veil-ripping, new technology of ours was reconfiguring our world as alien. A giant ant is no longer an ant, it is an alien invader.

'Since the coming of the atomic age,' ran the sonorous introduction to the movie *It Came from Beneath the Sea* (Robert Gordon, 1955), 'man's knowledge has so increased that any upheaval of nature would not be beyond his belief.'

The plot of the film concerns a sea monster that menaces San Francisco. Its normal habitat was the Mindanao Deep, but it had been disturbed by hydrogen bomb tests in the Marshall Islands. It was not only extraterrestrial aliens that were indignant at our meddling.

Note that they were hydrogen bombs. As if atom bombs weren't disorientating enough, the United States had exploded its first H-bomb in November 1952. It was codenamed Mike and it yielded a blast equivalent to 10.4 million tons of TNT, greater than all the ordnance detonated in World Wars I and II combined – and you thought they were bad news. The Soviets had acquired

the A-bomb in 1949 and, soon after Mike, they also acquired the
H-bomb. We were rapidly shaping up for a real 'war to end all
wars'.

Things were now well 'beyond belief', they were running out
of control. Indeed, one of the oddities of H-bombs was that they
seemed to be intrinsically out of control. If A-bombs were
cautious surrealists, H-bombs were unpredictable Dadaists.
Exact yields of these monsters could not be calculated in advance.
At 15 megatons, for example, the Bravo bomb of 1954 was 2.5
times bigger than the scientists had predicted. This technology
had a mind of its own. This technology was alien even to its
inventors.

So was the new strategic climate, though, in fact, this did not
seem to have any inventors. It was a human creation but it
happened without anybody actually having to do anything. The
Cold War, which the world was to fight until 1989, was not
something which we did, it was something done to us. In March
1946 at Fulton, Missouri, Winston Churchill had announced the
beginning of this ghostly cryo-conflict.

'From Stettin in the Baltic to Trieste in the Adriatic an iron
curtain has descended across the Continent. Behind that line lie
all the capitals of the ancient states of Central and Eastern
Europe. Warsaw, Berlin, Prague, Vienna, Budapest, Belgrade,
Bucharest and Sofia; all these famous cities and the populations
around them lie in what I must call the Soviet sphere, and all are
subject, in one form or another, not only to Soviet influence but
to a very high and in some cases increasing measure of control
from Moscow.'

The litany of cities was Churchillian poetry, but it was also a
target list. The Soviets did not then have the Bomb; the
Americans did and they could pick off these places like a row of
ducks. The Cold War meant that not only were the bombs here,
there was also a reason to use them. With Nazism defeated,
Communism had risen to take its place. Whatever horrors we
inflicted on each other, nothing in earthbound politics ever
seemed to change. No wonder Klaatu was frustrated in his
attempt to bring our leaders together. The very fact that Robert
Wise included that point in his film indicates a certain resigned

cynicism about the human ability to construct peace. And, if
there was no such ability, then the bombs would surely fall.

But, if politics hadn't changed, strategy had. The strategists
had no choice; this was a surreal world that required a surreal
response. Nobody had yet worked out how to fight a nuclear war
– in fact, luckily, nobody ever has – and, as they struggled with
these new concepts, they were constantly being distracted by the
saucers. It is important to note here that, though the
extraterrestrial hypothesis was not widely accepted, the reality of
the saucers was. Few people dismissed them as visions or
hallucinations and, on the whole, the American military certainly
didn't. But what were they?

Perhaps they were Soviet craft. A rather stunningly ingenious
explanation came from Naval Intelligence in 1948. The Soviets
were deploying the saucers 'to negate US confidence in the atom
bomb as the most advanced and decisive weapon', The objects
had first appeared as the 'ghost rockets' over Scandinavia, close
to the Soviet border. They appeared at the time that the US 'was
making a vigorous campaign for the economic and political
alignment of these nations with other pro-American Western
European nations'. Then they appeared over the United States.
The obvious conclusion was that the Soviets were using the craft
to convince people that, in spite of their lack of an atom bomb,
some of their technology was far ahead of the Americans'. It
hadn't worked because most people had perversely assumed they
were secret American projects. The next step, mused the navy,
was for the Soviets to admit the saucers were not only real but
also Russian.

Weird as this may seem, the point was that, in this apocalyptic
climate, the suspicion quickly spread that the Cold War, the
bombs and the UFOs were more than just coincidentally linked.
In part, this was just a matter of imagery. Towards the end of
World War II the Germans had launched V-2 rockets against
Britain. They didn't do much damage but they were the first
portent of what was, technologically, to come.

They were incredible things. Many British scientists hadn't
believed such a device could work at all and, when they did, they
arrived faster than the speed of sound, having first risen out of the

atmosphere. Then, after they landed, there was, as well as the explosion, this terrible rushing boom as the air refilled the tubular vacuum they had created. And they looked exactly like rockets should – a Groucho Marx cigar with one pointed end and fins. The SF illustrators had been right all along.

The cold warriors grabbed the rockets. From 1946 onwards 'our Germans' – the scientists abducted by the Americans – launched sixty-three V-2 rockets. In June 1948 at White Sands Proving Grounds, New Mexico, Werner von Braun, the leading German rocket scientist, launched a V-2 that took monkeys into the upper atmosphere. There were living things almost in space and we had put them there. Meanwhile, the Russians set about restoring V-2 production in its own territories. But the US was well ahead, having stolen more and better Germans.

'We defeated Nazi armies,' moaned Stalin, prior, presumably, to having someone shot, 'we occupied Berlin and Peenemünde; but the Americans got the rocket engineers. What could be more revolting and inexcusable? How and why was this allowed to happen?'

The SF imagery of the V-2s combined with the flying saucers to suggest that the space age and nuclear annihilation were connected. The bombs and the aliens were imaginatively linked. But was there also a moral or strategic linkage?

'Could the phenomenon itself,' asks David Darlington in his book *The Dreamland Chronicles: The Legends of Area 51 – America's Most Secret Military Base*, 'have been physically triggered by atomic explosions? Was some kind of metaphysical intelligence disturbed by the path that *Homo sapiens* was taking? Might actual extraterrestrial beings have been attracted to Earth by sudden flashes appearing over interstellar distances? Or was the collective *terrestrial* imagination simply so traumatized by the technopolitics of the time that it responded hysterically, conjuring visions of alien intruders?'

The possibility that the aliens had noticed our bombs struck many at the time. There was a strange, indeed sensational, appendix to the report of Project Sign, published in April 1949. This was in the form of a letter from James E. Lipp of the Air Materiel Command's Missiles Division to AMC's director of

research. Lipp concocts a plausible scenario to explain why the
saucers had come.

Martians have kept a long-term routine watch on Earth and
have been alarmed by the sight of our A-bomb shots as
evidence that we are warlike and on the threshold of space
travel. (Venus is eliminated because her cloudy atmosphere
would make such a survey impractical.) The first flying
objects were spotted in the spring of 1947, after a total of 5
atomic explosions, i.e., Alamogordo, Hiroshima, Nagasaki,
Crossroads A and Crossroads B. Of these, the first two were
in positions to be seen from Mars, the third was very
doubtful (at the edge of the Earth's disc in daylight) and the
last two were on the wrong side of the Earth. It is likely that
Martian astronomers, with their thin atmosphere, could
build telescopes big enough to see A-bomb explosions on
Earth, even though we were 165 and 153 million miles
away, respectively, on the Alamogordo and Hiroshima
dates.

The intro to the *Roswell Daily Record* story and Lipp's
appendix have one important thing in common. Neither could
now be written. We know there is nobody with a telescope on
Mars and we either know there are no flying saucers or we know
that there are and the evidence of their presence is systematically
concealed.

But anything was possible in the frenzy of 1947. And the truth
was we couldn't cope. For the message of the Bomb was that, in
spite of washing machines, bikinis and transistors, in spite of the
CIA, our rationality was inadequate. It had driven us mad.
'MAD', in fact, was the Cold War acronym for the strategy of the
age. It meant Mutually Assured Destruction. We could only keep
the peace by threatening to kill everybody. It made sense, but,
then again, it didn't. Why not just have peace and threaten
nobody? No, that didn't make sense either.

What was needed was a higher rationality, an alien rationality.
Gort with his implacable beam represented such a higher
rationality. He was deliberately created to be beyond strategy and

negotiation, he simply stopped conflict. There was no need for an inner Gort, a reflective Gort, because there was nothing to think about. Or perhaps there was an inner Gort but his utterly resolved thought processes were beyond our comprehension. Robby, however, has an inner life. He is the robot in *Forbidden Planet* (Fred McLeod Wilcox, 1956). He is a dumpy creature, still stuck in the heavy age of industry rather than the coming, weightless age of information. He is chubby, bristling with aerials and three gyroscopes roll around inside his head. On the planet Altair 4 he looks after his creator, Dr Morbius, and his daughter, Altaira, 'like a mother'.

Robby cannot harm a rational being, such an act would conflict with his programming. Morbius demonstrates this product feature to some interplanetary visitors by ordering the robot to kill a man. Robby goes into a form of epileptic seizure in which sparks fly around inside his transparent headpiece. He wishes to obey the order, but his programming prevents him. Luckily, Robby is merely paralysed by this conflict; twenty-two years later a similar crisis will render Hal, the computer in *2001: A Space Odyssey* (Stanley Kubrick, 1968), homicidally insane. Robby does, however, look upset when a crisis occurs, though Morbius insists he is without emotion.

Forbidden Planet is based on Shakespeare's *The Tempest* and Robby is the technological version of the good sprite Ariel. His counterpart, the bad monster Caliban, is an aggressive, usually invisible creature that can occasionally be seen as a gigantic, snarling, bear-like thing, as when he is electrically shocked into visibility by a force-field fence erected by soldiers.

This Caliban, we assume, is an inhabitant of the planet. But he is not. He is pure, Freudian id, a projection of 'the beast, the mindless primitive' that lurks within Morbius's subconscious. Morbius, unconsciously, has made him.

This idea of the projection into the real world of a psychic condition is important. It bridges the gap between mind and matter by implying that we can think something into material existence. Mind is therefore implicated in matter and matter in mind. The strong version of this idea is that our minds can project things that would be objectively real to others; the weaker version is that we project things that are only real to ourselves. In various forms, these possibilities repeatedly appear in alienology.

Psychic projection is important, at one level, as an explanation, the one favoured by Jung, of the UFO phenomenon. Jung, however, was not clear – he seldom is – about the reality of these projections, whether one person's projection could be seen by another. But, specifically in the context of *Forbidden Planet*, the idea symbolizes an anxiety about both the power and malleability of the human mind. Morbius projected his Caliban, but he did so unconsciously. It was an aspect of him over which he had no control. This relates directly to 'brainwashing', one of the Cold War fears in the air at the time. This was a possible technique whereby people could, in effect, be turned into somebody else or programmed, like a computer, to perform a particular task. Technology, in other words, could make people into aliens. *The Manchurian Candidate* (John Frankenheimer, 1962) demonstrated its potential – a soldier brainwashed during the Korean War returns to the US as an assassin, a man possessed of a Gort-like, logical implacability. Morbius's projection of the monster signifies both the human evil that lurks in the paradise of the Forbidden Planet and the widespread belief at the time that the mind had as yet unknown capabilities . . . and vulnerabilities.

The revelation of the monster's true nature comes via Robby/Ariel. He is ordered to defend the humans against this Caliban. Anxious sparks fly around his headpiece. He can't kill the beast because he recognizes that it is an aspect of his master, Morbius. Again the alien, this time a machine, is better than we are. Of course, Robby is simply programmed by humans to be better; like Gort he is primed with an unbreakable moral absolute. But the point is he *can* be programmed. We can't. Or, at least, we can't any more.

'We've all got monsters in our subconscious,' says the hero who sees the beast for what it is, 'so we have laws and religion.'

Laws and religion are our programmers. But with God already a long time dead and the hippies and postmodernity barely a decade in the future, in 1956 laws and religion were pretty clearly on the way out. Besides, neither laws nor religion had done a very good job of programming us. This film was made only eleven

years after the first pictures of the concentration camps had been
seen and after Hiroshima and Nagasaki. The beast had got out of
the subconscious. In truth, he had never spent much time there
anyway, he'd always preferred the fun of being outside in the full
light of consciousness where he could do the most harm. But,
now, of course, he could do so much more.

Robby, like Gort, dramatizes human shortcomings that,
through the prosthetic extensions of rockets and atom bombs,
were about to become very dangerous indeed. It was bad enough
that war in the twentieth century had been expanded to include
the massacre of civilian populations. But these atomic bombs
meant the whole world, if not the entire galaxy, could be a
'legitimate' target. Technology could offer labour-saving gadgets
in bright new homes; equally, however, it could offer apocalypse
at the flick of a switch. No wonder the aliens were on the case.

At first, people, as opposed to the media and Hollywood, did
not, in general, associate UFOs with aliens. So these warnings
about human evil tended to be fictional. But the technological
acceleration meant that the most common habitat of the aliens –
science fiction – was struggling to keep up. The futures of pre-war
SF were always just futures, now, suddenly, they were imminent
possibilities. When the Soviets launched the first satellite,
Sputnik, in 1957, the whole genre seemed under threat.

'There may have been a time, in the morning of the world,'
wrote critic John Clute, 'before Sputnik, when the empires of our
SF dreams were governed according to rules neatly written out in
the pages of Astounding and we could all play the game of a
future we all shared, readers, writers, fans . . . But something
happened. The future began to come true.'

Sputnik was a real human thing in space. We really were out
there and, if things carried on at this pace, the interstellar cruisers
could only be a few years away. The future was, indeed, coming
true. But aliens had also been in the SF future. They must be on
their way too. They were. Reports of actual encounters with
aliens and, subsequently, of abductions by aliens began to
appear. These real-world extraterrestrials started to deliver the
same warnings as their fictional precursors. In 1967, for example,
Ludwig Pallman, a German businessman, was taken by aliens to

a base in the Amazonian jungle. The aliens told him they had received specific orders with regard to Earth.

'We are here because we know about the calamities that will overrun Planet Earth – one of them, the Third World War. It is not far away; we know the date but we cannot interfere . . . You must learn to live in peace and harmony, but you have to do it by yourselves, and that implies gaining a new level of conscience.'

Of course, there is nothing novel about the end of the world. The Christian day of judgment had always been just a heartbeat away. Amidst secularity and growing affluence, that form of the anxiety may have faded somewhat. But the anxiety itself burns as darkly as ever, only the mechanism has changed. In the immediate postwar world, that mechanism was the human capacity for violence extended by technology. Later it was to be further extended by our capacity for environmental destruction.

In 1958 Jung published his *Flying Saucers: A Modern Myth of Things Seen in the Sky*. Jung viewed the saucer sightings as an opportunity to observe a myth in the process of formation. He saw the saucers themselves as psychic projections – placed in the sky by our own fevered consciousness, though perhaps no less real for that. The question was: why should we be projecting such things at this particular moment? The answer was that it was a moment fraught with a peculiarly novel anxiety.

'In the threatening situation of the world today,' Jung wrote, 'when people are beginning to see that everything is at stake, the projection-creating fantasy soars beyond the realm of earthly organizations and powers into the heavens, into interstellar space, where the rulers of human fate, the gods, once had their abode in the planets.'

And he added: 'The basis for this kind of rumour is an *emotional tension* having its cause in a situation of collective distress or danger, or in a vital psychic need.'

Jacques Vallée, UFO researcher and the model for the French scientist Claude Lacombe played by François Truffaut in the film *Close Encounters of the Third Kind* (Steven Spielberg, 1977), made a similar point after studying thousands of accounts of UFO and alien experiences, writing that 'such accounts appear to be a way for certain souls to release their anguish in the face of modern scientific changes, their fear of war and atomic cataclysm and their inability to adapt to the present rhythm of life'.

If, of course, the aliens in their UFOs are psychic projections or occasions of released anguish, then they are not literally extraterrestrial beings. Kenneth Arnold had seen his saucers, but he had also created them, projecting them against the snows of Mount Rainier. The UFOs are aspects of a myth created under pressure in the same way that religions might be created in fear of death or incomprehension at the spectacle of human suffering. They are, as author Mike Jay puts it, 'a kind of folk psychiatry', a therapy session conducted to alleviate a global neurosis. This seems to be the logical explanation of a somewhat glaring coincidence – the birth of ufology and the contemporary wave of alien encounters just at the moment we acquired rockets and atom bombs. But there could be another, equally logical, nuts and bolts explanation. The aliens are literally here and they're here because they don't like the bombs.

Technology had staged a re-enactment of the fall of man and we either required – the psychic projection theory – or we were being offered – the real aliens theory – a new salvation. In the late forties and the fifties, they were here to save us from our warlike stupidity. But, in the early sixties, a new human crime was exposed which was to become, thereafter, the dominant cause of alien indignation.

In 1962 Rachel Carson, a naturalist, zoologist and already a successful writer, published *Silent Spring*. She was fifty-five at the time and was to die two years later. The book created the modern environmental movement. Its key insight was that the science behind the tons of pesticides and herbicides being used at the time was wrong. Although the amounts being applied seemed harmless, the effect of the food chain was to concentrate these poisons in certain species. It was the destruction of the songbird population because of the indiscriminate use of DDT that gave Carson her title. Reviewing the book, the *New York Times* specifically linked the insight to the nuclear threat.

'She fears,' wrote Lorus and Margery Milne, 'the insidious poisons, spread as sprays and dust or put in foods, far more than the radioactive debris from a nuclear war.'

The poisons were, like radiation, invisible killers. But it was the totality of the environmental threat that most closely linked these chemicals to the Bomb. Once again we had discovered a new human power to destroy everything.

'Man's attitude toward nature,' said Carson in a CBS interview shortly before her death, 'is today critically important simply because we have now acquired a fateful power to alter and destroy nature. But man is a part of nature, and his war against nature is inevitably a war against himself . . . [We are] challenged as mankind has never been challenged.'

The threat was unprecedented. Before, we just farmed, but now we were dosing the entire landscape. Furthermore, we were spreading like vermin. Over-population was another of the primary fears of the postwar years. In 'The People Trap', a short story of 1968, Robert Sheckley defined this gross folly of our age from the perspective of the future.

'The seeds of the problem had been sown in the twentieth century; but the terrible harvest was reaped a hundred years later. After uncounted millennia of slow increase, the population of the world suddenly exploded, doubled, and doubled again. With disease checked and food supplies assured, death rates continued to fall as birthrates rose. Caught in a nightmare geometric progression, the ranks of humanity swelled like runaway cancers.'

We had become a disease. The Bomb, over-population, consumption, the environment all said the same thing. There had been a phase shift in our activities. Klaatu had not been concerned when we just had tanks and bombs, but worried once we went nuclear. Similarly, thousands of aliens were now to emerge, all alarmed that our meddling and greed had got out of hand.

'Your world is marvellous,' Julion Fernandez, a Spanish businessman, was told telepathically by aliens in 1978, 'its biological richness is unbelievable. There are very few worlds like it. We ourselves don't know of a single other one. It is a veritable mine, a well-nigh inexhaustible mine of many of the things that we need, and that we do not have: water for one. Unfortunately you yourselves have already begun the process of destruction of this world. It is a story that has been repeated on many other occasions: it happened in our own worlds once, centuries ago . . .'

'Many abductees recall having been shown visions of a dying planet,' writes abductee investigator and counsellor Budd Hopkins, 'harrowing 3-D images of earthquakes and fire and nuclear destruction.'

Hopkins exhaustively investigated the Brooklyn Bridge abduction case and concluded it was a deliberate attempt by the aliens to announce their presence. Three aliens (of the classic 'Grey' type, the one most commonly encountered) had extracted a woman from her Manhattan flat and levitated her into a saucer. It was, apparently, witnessed by the secretary-general of the United Nations, Javier Pérez de Cuellar, though he never confirmed this and Hopkins concealed his name. The abductee, Linda Cortile (a pseudonym), was used by her captors to deliver angry environmental harangues: '. . . it seems that whenever they need her, the aliens can throw a mental switch, as it were, transforming her into a fire-breathing environmentalist, railing at her fellow human beings, waving dead fish and effortlessly speaking an alien language'.

Our environmental crimes tell us something, but it is something we knew already, something we had always known. We are not right in this world. Unlike animals, we cannot just be, we must meddle, we must get and spend. Our meddling and our greed are selfish and short-sighted. We are an offence to nature and, it transpires, the extraterrestrials. Or perhaps the extraterrestrials *are* nature, woodland sprites or goblins, sent to bring us back to Earth, our home.

Alienology is riddled with a sense of original sin, with a sense not just that we are making a mistake, but that we can only make mistakes. Usually it is good aliens that point this out, in the hope that we shall reform ourselves. But sometimes, the bad ones do as well. In *The Matrix* (Andy and Larry Wachowski, 1999), Agent Smith, the computer program created by the machines to quell human dissidence, tells the human Morpheus what he thinks of us.

I'd like to share a revelation that I've had during my time here. It came to me when I tried to classify your species. I realized that you're not actually mammals. Every mammal on this planet instinctively develops a natural equilibrium with the surrounding environment, but you humans do not. You move to an area, and you multiply, and multiply, until every natural resource is consumed. The only way you can survive is to spread to another area. There is another organism on this planet that follows the same pattern. A virus. Human beings are a disease, a cancer of this planet, you are a plague, and we are the cure.

Smith says this while disgustedly touching the sweat on Morpheus's face. It is a remarkable speech, an entirely legitimate defence of the position of the machines. Contrast this with the aspiration of *Star Trek*'s android Data to become human. This theme, threaded through countless episodes and several films, is based on the assumption that the highest destiny for a machine is to increase its humanity. But this is a legacy of the humanist roots of *Star Trek* in the sixties. In contrast, Smith the computer program, a virtual machine, aligns himself with the environmental movement of the nineties in finding humanity a stain upon the Earth. We are disgusting, we are out of step with nature, we are like some virulent disease. Let the aliens – machines in this case – take over. Agent Smith is pure software, he does not sweat. He is free of the restraints and filth of corporeality. The very existence of the digital world is a rebuke to messy humanity. And, whether we like it nor not, we are entering a digital world.

'The analog world,' writes Erik Davis in his book *TechGnosis: Myth, Magic and Mysticism in the Age of Information*, 'sticks to the grooves of the soul – warm, undulating, worn with the pops and scratches of metahistory. The digital world boots up with the cool matrix of the spirit: luminous, abstract, more code than corporeality.'

Davis implicitly points to the next phase in the alien invasion – the phase of the spirit. Software is to hardware as the spirit is to the body. And our crimes – war, over-population, ecological destruction – are the crimes of the body committed by excessively embodied people. His Holiness the Dalai Lama saw the point. In 1992 he said aliens were sentient beings and 'apparently troubled by what is happening on Earth'. Increasingly, alien lore began to focus not just on our malign physics or our murderous agriculture, but on our spiritual failings.

The aliens who warned Ludwig Pallman about the Third World War also told him: 'Perhaps the greatest discovery that Earthlings will make, will be to know God . . . You believe in a completely mistaken concept of God.' The alien abductee Karen told the Harvard Professor of Psychiatry John Mack: 'We've written God completely out of our awareness.' And Mack himself commented: 'The secular modern, or postmodern, Western culture in which I was raised and have lived is perhaps unique in the degree to which many of its members have become separated from any sense of the actuality of divine presence or higher power.'

American philosopher Michael Zimmerman makes a similar point.

> Unwilling to live in this spiritual desert, countless people are reporting encounters with extraordinary realms and entities long regarded as the products of primitive imagination. Instead of concluding that near-death experiences, past-life regressions, encounters with angels, and alien abductions are instances of escapism, delusional thinking, irrationality, or psychosis, we might entertain the possibility – as did William James – that these phenomena reveal dimensions of reality that are hidden to ordinary consciousness, and that invite spiritual and personal development on the part of those who encounter them.

Thomas E. Bullard points out that millenarian movements throughout history have sprung 'from the need to revitalize a society shattered by deprivation and despair'. Modern society, with its disappointed faith in human science, may thus be shattered. 'A dose of revitalization seems in order for the modern secular condition,' suggests Bullard. The aliens offer that dose.

The source of the alien indignation has been expanded massively. It is no longer merely nuclear weapons or environmental destruction; it is now dismay at our spiritual depravity. Furthermore, it is a critique of Western scientific and technological culture. Our narrow, empirical focus has blinded us to the great spiritual truths of the universe. It is an idea that has possessed those who have encountered aliens just as it possessed earlier generations who turned to the occult to expose the banality and limitations of the world view of science. But it is also an idea that can be contained within the focus of science. The physicist Paul Davies acknowledges the sense of emptiness bequeathed to us by the scientific world view, but believes the discovery of aliens, far from humiliating us, could lead to a new sense of human purpose.

'In my opinion,' he writes, 'the most important upshot of the discovery of extraterrestrial life would be to restore to human beings something of the dignity of which science has robbed them. Far from exposing *Homo sapiens* as an inferior creature in the vast cosmos, the certain existence of alien beings would give us cause to believe that we, in our humble way, were a part of a larger, majestic process of cosmic self-knowledge.'

This is not so far from the messages delivered by interstellar beings to cult groups or channelled through mediums. These repeatedly tell us that we have wandered into a spiritual desert. Indeed, the Nine, a committee of intergalactic, interdimensional beings who channel their communications through the Florida medium Phyllis V. Schlemmer, say that our obduracy is blocking the evolution to a higher plane of the entire cosmos. But there is hope.

'Man is now coming out of the true dark ages of the planet,' says Tom, the spokesman of the Nine, 'and becoming aware of the existence of other life forms in other parts of the universe. Men have always assumed there was something sitting up there taking care of their problems, but they also assumed through their ego that they were the only existence and that this being called God was only concerned with them. Man has now to understand that there are other forms of life and that the universe does not revolve, or evolve, just around man.'

Tom's point is highly significant. Man wants to be the summit of creation. Either he needs a God to tell him that this is, indeed, the case, or perhaps he prefers to tell himself in the absence of God. The problem with both positions is that we seem to be a pretty poor summit. God may regard us as his finest work, He may have constructed the entire cosmos around us. But we still spend our days devising new ways of killing each other and befouling our home. Christianity explained this through original sin. We used our freedom to make a choice, in the event the wrong one. But we are held back from a complete descent into evil by the prospect of judgment and damnation as well as the possibility of salvation. In the climate of the postwar world, this seemed to have become inadequate. The threat of atomic or environmental destruction implied that the judgment of God was no longer powerful enough to hold us back. And so the aliens arrived with their superior technology to thwart our pretensions.

Their arrival also announces that we are not the summit of creation. We have been dethroned. Sinners that we are, our first instinct seems to have been to seize this extraterrestrial technology to deploy in our terrestrial wars. But that only works if the aliens want to give it to us. Obviously, if they really are so superior, then we cannot take it from them. Gort, the implacable robot, makes the point. We pull our guns and he melts them. So the indignant aliens are obliged to humiliate us, to tell us that we are not only not on the throne, we are not even fit to join the cosmic community. Out there something wonderful is happening and human beings are simply not good enough to take part.

Olaf Stapledon published his novel *Star Maker* in 1937. It anticipated the postwar idea that life on Earth was but a trivial sideshow in the vast fairground of the cosmos. Stapledon's hero walks out of his house one night, stricken with 'horror at our futility, at our own unreality.' He climbs a hill. The sight of the stars strikes fear into his heart, 'For in such a universe as this what significance could there be in our fortuitous, our frail, our evanescent community?' Mysteriously, he is then transported out into the heavens. He looks back at Earth and is struck, as later, real astronauts would be, by its fragile beauty. He feels that it is 'a creature alive but tranced and obscurely yearning to wake'. He notices that he can see no evidence of life.

'No visiting angel, or explorer from another planet, could have guessed that this bland orb teemed with vermin, with world-mastering, self-torturing, incipiently angelic beasts.'

What follows is a voyage through a cosmos teeming with life and higher and higher levels of consciousness; even the stars, it transpires, are conscious. Far from being the empty wasteland human science had constructed, it is a fabulous drama from which our pettiness has excluded us. So far beyond human consciousness is the activity of the universe that Stapledon's traveller cannot accurately recount its nature in our terms. But he attempts a summary.

The main practical work was to enrich and harmonize the life of the galaxy itself, to increase the number and diversity and mental unity of the fully awakened world up to the point which, it was believed, was demanded for the emergence of a mode of experience more awakened than any hitherto attained. The second kind of activity was that which sought to make closer contact with the other galaxies by physical and telepathic study. The third was the spiritual exercise appropriate to beings of the rank of the world-minds. This last seems to have been concerned (or will be concerned) at once with the deepening of the self-awareness of each individual world-spirit and the detachment of its will from merely private fulfilment.

Perhaps the coming of the saucers was the first step in our initiation into these wonders. Perhaps, in the words of Donald Keyhoe, the most influential early investigator of the UFOs, we 'stood on the threshold of something strange and tremendous'. Judging by the anger of the aliens, this first step involved a cosmic reprimand for our violent folly.

These angry aliens who began to arrive in 1947, buzzing our military installations, were a warning that the price we were paying for affluence was too high. The bikini always came with a bomb attached. Beneath the good life, we were the same old human beings – warring, dirty, trivial. The shining light of goodness that Helen Benson sees in Klaatu's eyes will always contrast with the rat-like gaze of her cheap grifter of a human boyfriend. And yet, as Stapledon's hero acknowledges from the depths of his bitter view of humanity, we are 'incipiently angelic'.

The aliens may be here to improve us, to seize our angelic natures, to lift us from the mire of human history.

But why, in that case, were they just flying about the place? Why were they so elusive? Why weren't they making contact as they did in the movies? Why weren't they talking to us? In fact, they were.

Betty and Barney Hill

In 1966 *Look* magazine published two excerpts from a book by
John G. Fuller entitled *The Interrupted Journey*. The book was
about Betty and Barney Hill. Barney was a 39-year-old mail
sorter and Betty a 41-year-old social worker. They were a
respectable couple. Their only oddity, for the time, was that he
was black and she was white. They were both active in their
church and in the civil rights movement. They seemed to be
normal, stable, grounded. Like Kenneth Arnold, they were good
witness material.

The book was about an incident that occurred at 10pm on
19 September 1961. The Hills were driving south on US Highway
3, returning from a holiday in Montreal to their home in
Portsmouth, New Hampshire. They noticed something that
looked like a star, but which seemed to be following them.
Through a pair of binoculars, Betty saw the object had double
windows. Finally, when the object was only 50 feet away from
them, Barney stopped the car. They were near Indian Head in
New Hampshire. They both got out of the car.

Barney described the object as a large glowing pancake and
assumed it was some kind of military craft, though he found it
odd that it should stop and observe them. In fact, he saw there

were several figures watching them from the windows and became fearful at the possibility that they planned to kill them both for having seen this top-secret craft. They got back into the car and drove quickly away.

Next they heard a beeping noise from the boot of their car which seemed to make them tired. The beeping sounded again and they returned to full wakefulness, finding themselves some miles to the south of where the incident had happened. They continued home. But they arrived two and a half hours later than they should have done. They subsequently found a dozen or more shiny circles scattered around the boot of the car and they reported the incident to Pease Air Force Base in Portsmouth. Later it was to emerge that Strategic Air Command's 100th Bomb Wing had registered an unidentified object on the Pease radar at the same time as the Hills had encountered the strange craft.

Betty Hill began to have nightmares in which she was taken on board a UFO and examined by small humanoids. After a doctor had checked them out, they were both referred to Dr Benjamin Simon, a psychiatrist and neurologist, who hypnotized them. Under hypnosis, they told identical stories about what had happened in those hours of missing time. Simon was never to believe they had been abducted, eventually concluding they had experienced some sort of shared dream.

These were the details that emerged. The Hills' car had stalled, they were at once surrounded by humanoids in black uniforms and taken on board the craft. A needle was inserted into Betty's stomach and a circular device was attached to Barney's groin. Semen was extracted, though this emerged much later than the rest of the story as Barney had been embarrassed. Betty was given a tour of the craft and shown a star map. They were told they would remember nothing and returned to their car. The UFO then flew away. The story was later made into a TV movie – *The UFO Incident* (Richard A. Colla, 1975).

Like the Arnold sighting, this is often seen as the first incident of its kind. But, again, this is not true. In May 1951 Orfeo Angelucci was addressed by aliens on his way home from work in Burbank, California. The following year he was taken on board an alien craft. In October 1957 in Minas Gerais in Brazil,

Antonio Villas Boas was ploughing his family's field. It was one o'clock in the morning. He saw what appeared to be a red star approaching. The tractor stalled. A UFO landed nearby. It was a long, egg-shaped object supported by three legs. Villas Boas was surrounded by four small figures who dragged him inside the ship. Five more creatures stripped him, took a blood sample from beneath his chin and swabbed him with some kind of liquid. He was left in the room, which suddenly filled with grey smoke that caused him to vomit. Then a naked woman appeared who looked roughly human, but not quite. Villas Boas was uncontrollably aroused and they had sex twice, though he was slightly put off by the noises she made.

'Some of the growls that came from her . . . Gave me the disagreeable impression of lying with an animal.'

As the woman left, she pointed to her stomach and then to the sky, apparently indicating that he had impregnated her and their child would be brought up away from Earth.

The story was so fantastic that the original investigators were reluctant to go public and it was not to emerge fully until January 1965 in the British journal *Flying Saucer Review*. This was four years after the Hills' experience and the timing makes it clear that the similarities between the two cases – the descending bright object, the stalled motors, the small creatures and the medical procedures – could not have arisen from the Hills copying Villas Boas.

There were many other incidents. On 18 April 1961 in Wisconsin Joe Simonton saw a UFO land outside his house. He looked inside the craft and saw three crew members who looked Italian. Through gestures, they indicated they wanted him to fill a jug with water and, in return, they gave him four pancakes. The story was dismissed by many as absurd but it led to many more contact cases. On 24 April 1964, for example, Gary Wilcox, a farmer in Newark Valley, New York, found an egg-shaped object on his property. Inside were four four-foot tall humanoids who engaged him in conversation about, among other things, manure. The beings said they were from Mars, though it crossed Wilcox's mind that they might be from *Candid Camera*. In the end, as the writer Jerome Clark observes,

'Though Wilcox's account was unbelievable, there seemed no doubt that *he* believed it.'

In general, contact accounts are greeted with more scepticism than mere UFO sightings. Partly, this was because they simply seemed more incredible, but also they had a chequered history. George Adamski was part of the problem.

'I am George Adamski,' he introduced himself in *Flying Saucers Have Landed*, a book he co-wrote with Desmond Leslie, 'philosopher, student, teacher, saucer researcher'.

He was, in fact, a charming Polish-American hamburger-stand cook. He was already known in Californian esoteric circles for the Royal Order of Tibet he had founded in the 1930s. Adamski had not been content simply to see UFOs; he wanted to meet and travel with aliens. And so, at 12.30pm on Thursday 20 November, 1952, in the California desert he met Orthon, a long-haired young man from Venus. They communicated telepathically. Adamski also took a famous picture of the Venusian 'scout ship'. His descriptions of Orthon were detailed and odd.

'(1) His trousers were not like mine. They were in style, much like ski trousers and with a passing thought I wondered why he wore such out here in the desert.

'(2) His hair was long, reaching to his shoulders, and was blowing in the wind as was mine. But this was not too strange for I have seen a number of men who wore their hair almost that long.'

Adamski reckoned Orthon was about twenty-eight years old, 5'6" tall and weighed 135 lbs. His shoes were 'ox-blood in colour' and 'like a man's oxford'. They were, he estimated, about size nine or nine and a half. But, above all, the Venusian was friendly – his presence was 'like the warm embrace of great love and understanding wisdom'.

Pictures later drawn of Orthon make him look not unlike Klaatu and very similar to the alien encountered by amateur astronomer and ornithologist Cedric Allingham in February 1954 near Lossiemouth in England. This was a 'man' about six feet tall, who was basically humanoid but for his very high forehead. His clothes made a similarly strong impression on Allingham. He wore a one-piece garb with footwear incorporated, but no helmet.

Adamski subsequently flew round the moon with the aliens and visited Venus where his dead wife had been reincarnated. He died, perhaps going to join her, in 1965. Certainly he lives on here on Earth. As the Adamski Foundation web site www.gafintladamski.com explains, 'Through his devotion and courage to speak, he personally became responsible for pioneering the movement towards establishing greater public awareness and education regarding the existence of extraterrestrial life.' Or as Colin Bennett puts it in his book on Adamski entitled *Looking for Orthon*: 'Like many with a streak of genius, he didn't really know the difference between work and play, dream and religious impulse, inspiration and rational thought. But his faulty intellectual grasp saved him: it allowed him to play with all these things, and in playing he chanced upon something that talked to him.'

Adamski was definitely nuts and bolts, the UFOs and aliens were real aspects of our world. His contemporary George W. Van Tassel, however, veered more towards the third realm interpretation. In 1947, after a career in aviation, Van Tassel moved with his family to Giant Rock in the Mojave Desert in California. The place was named after a huge, freestanding boulder which had formerly been sacred to Native Americans. Van Tassel seems to have been in psychic contact with aliens from January 1952. In August 1953, Venusians landed and invited him on to their ship. They taught him a technique for rejuvenating living tissues and, the following year, he began building the Integratron – a 'high-voltage electrostatic generator that would supply a broad range of features to recharge the cell structure'. From 1954 until 1977 conventions were held at Giant Rock for contactees and communicators.

Jerome Clark says that Van Tassel was even more influential than Adamski because his conferences 'provided a congenial atmosphere for space communicants and their followers (known as "saucerians"), and in that environment a theology for the age of flying saucers developed, based on a complex (if less than entirely consistent) cosmology'.

To their various admirers, such men as Van Tassel and Adamski were either visionaries who had glimpsed the truth or culturally decisive fabulists, the godfathers of *Star Trek* and *The X-Files*. But, to serious ufologists, they were complete pains. The extravagance of such claims undermined the ufologists' own work by making them look foolish.

'We had at that time,' says Budd Hopkins, 'Adamski and all those hoaxers and con-men and screwballs who took rides to Venus and discredited the idea that there was any kind of genuine thing.'

And Donald Keyhoe was so embarrassed when an employee gave membership cards for his National Investigations Committee on Aerial Phenomena (NICAP) to Adamski and six other contactees that he sacked her at once.

A rift had been created in alien investigation. Writing in 1957, Isabel L. Davis, a confirmed ufologist as opposed to saucerian from the group Civilian Saucer Investigation (CSI), anatomized the issue:

> All flying saucerdom is now divided into two irreconcilable groups. One group believes that human beings have had direct communication with extraterrestrial beings; the other group rejects all such reports as the products of conscious or unconscious fraud.
>
> The split between believers and skeptics is, and should be, a real and permanent one. For to the skeptics, flying saucers still deserve the name of UFOs – *Unidentified* Flying Objects. To the believers, on the other hand, thanks to the extensive information they claim to have received from their extraterrestrial friends, the saucers are no longer UFOs but IFOs – *Identified*, fully identified, Flying Objects. The two terms are mutually exclusive. An object cannot be identified and unidentified at the same time.

This was an odd posture that seemed to require of these phenomena only that they stay in their correct, human categories. Ufology aspired to scientific rigour, but, paradoxically, this meant that UFOs must remain mysterious and opaque, precluding any speculation or claims about what or who might be inside. It was as if simple observation of the craft – what

investigator J. Allen Hynek was to classify as Close Encounters of
the First or Second Kinds – was intrinsically more respectable,
more objective than actual interactions with their crew – Close
Encounters of the Third Kind. It was perhaps simply that
meetings and abductions asked too much of the averagely
sceptical layman. The battle for the hearts of the open-minded
would not be won by providing details of Orthon's shoes. As a
result, even when, after the Hill case, abduction reports started to
become almost as common as sightings, many ufologists refused
to have anything to do with them.

The credibility of the Hills thus became a crucial issue. Old
ufologists wanted theirs to be a false report; but a new wave
wanted it to be true. The most powerful discrediting weapon in
the sceptics' armoury was media influence. An episode of *The
Outer Limits* TV series entitled 'Bellero Shield' was broadcast less
than two weeks before Barney described the aliens he had met. It
is about mankind's first contact with an alien, who has wraparound eyes and is able to speak telepathically to the humans.
Barney's aliens had the same two qualities. Betty's account also
has similarities to scenes in the film *Invaders from Mars* (William
Cameron Menzies, 1953). Also *Killers from Space* (W. Lee
Wilder, 1954) could have been the inspiration for the 'amnesia
with recoverable memory' that appeared in the Hills' case and
thousands of subsequent abduction episodes.

Believers, on the other hand, point to the obvious sanity and
respectability of the couple and to the physical evidence. Barney's
shoes were scuffed – he had been forcibly dragged into the alien
craft – and a semi-circle of warts appeared around his groin. Pink
spots appeared on Betty's dress.

Either way, the net result of the Hill and Villas Boas cases was
the general dissemination of the idea of contact and abduction
from the mid-sixties onwards. This was a turning point. Arnold
had inaugurated the era in which the world became accustomed
to the idea of UFOs simply as enigmatic objects in the sky. That
these were alien craft was only one possible explanation among
many. But the Hill case was based upon the insistence that the
craft were indeed occupied and flown by aliens. As Budd
Hopkins put it to me: 'It took them twenty years to accept the

idea that UFOs had an inside.' And crucially, like Arnold, the Hills were just very ordinary people to whom something extraordinary had happened. The very banality of the circumstances – the highway at night, the car, the warts, the spots on the dress – indicated that this could happen to anyone, anywhere.

Abduction raised the stakes, for it meant that the aliens had more than just a passing interest in the human race. They were not content passively to observe, rather they had some active project that involved traumatic disruption to our lives. This could still mean their intentions were benign, that they were here to stop us destroying ourselves and the planet or to induct us into the cosmic community. But it introduced the possibility of something terrifying, of an alien takeover, a project to exploit, enslave or massacre humanity. And it was clear that, if the Hills were right, there was nothing whatsoever we could do about it.

Alien Surgery

Budd Hopkins was born seventy-three years ago in West Virginia. He fell in love with painting at an early age and, after attending lectures by the artist Robert Motherwell in 1952, he determined to move to New York City. There he became an important figure in the abstract expressionist movement that was to establish New York as the postwar world capital of the visual arts.

He remains a respected artist. But, latterly, he has become more well-known for his other work. For Hopkins is the central figure, a gatekeeper, in the vast and strange otherworld of alien abduction, the temporary removal of thousands, if not millions, of people by extraterrestrials for a series of imperfectly understood purposes ranging from simple curiosity to hybrid breeding programmes. Hopkins is, according to sociologist Ron Westrum, the man responsible for the 'paradigm shift' that took ufology from the analysis of simple sightings into the more psychologically fraught realm of abduction studies.

Since the Betty and Barney Hill revelations, alien abduction has become a distinct body of alienology with its own anthology of crucial cases. Two examples will suffice. In 1975 there was the Travis Walton abduction in Snowflake, Arizona, in which seven men spotted a UFO hovering above the trees. A beam of light from the craft struck Walton unconscious. The others fled in terror. When they returned to find Walton he had vanished. He reappeared five days later in a state of almost total amnesia. In 1976 there was the multiple abduction of a canoeing party on the Allagash River in Maine. Four people in a canoe were caught in a beam of light from a UFO. They then found they had 'lost' two or three hours. Under hypnotic regression they recalled being made to lie naked on benches and subjected to complex physical examinations as well as the removal of saliva, skin, blood, faeces, urine and semen. And so on.

There are thousands of such tales. Indeed, the phenomenon is so common that various companies have offered insurance against the trauma and possible physical injuries caused by abduction. However, in 1997 the London firm of Goodfellow Rebecca Ingrams Pearson withdrew from this particular market. Ten per cent of their business came from these policies, largely in Britain and the United States. But they pulled out because they found they had issued a policy covering up to fifty members of the Heaven's Gate cult in San Diego, thirty-nine of whom killed themselves in the sure expectation of being taken aboard a passing alien ship.

If you are denied insurance, however, you can still get cover. Michael Menken in Bellevue, Washington, offers an 'experimental device to stop alien abduction'. This 'thought screen helmet' blocks alien telepathy and mind control. It is made of leather with several layers of 'special conductive plastic'.

Abduction takes the alien issue to a new level, hardening the confrontation between sceptics and believers. Sceptics can scoff at the frequently comical absurdity of many of the tales. Believers can point to the extraordinary number of abductees telling remarkably consistent stories. Surely, they cannot all be insane.

Budd Hopkins is, indeed, a living rebuke to the kind of dogmatic scepticism that would portray all abductees and believers as mad. There is nothing weird or crazy about Hopkins. Indeed, he is an almost extravagantly consoling figure. He lives where he has lived for years – on West 16th Street – and he looks like an ad-man's dream of an East Coast grandfather – pullover, jeans, bushy eyebrows, silky white hair and the kind of healthy but crumpled features that have become an emblem of American wisdom and goodness. He has not been abducted himself, but he is the leading believer in abductions and he insists that very few of the abductees he has met are remotely disordered.

One of the enduring puzzles for sceptical psychologists who have studied alien abductees or experiencers in general is that they seldom show any signs of mental disturbance or personality disorder other than those directly associated with the experience.

'Most people who report these experiences display average to above average intelligence,' reports psychologist and neuro-scientist Michael Persinger, 'are not "crazy", and are very aware of the social and personal consequences of their experiences upon their families, friends, and vocational opportunities.'

Nicholas Spanos of Carleton University found 'no support whatsoever for the hypothesis that UFO reporters are psycho-logically disturbed . . . the onus is on those who favor the psychopathology hypothesis to provide support for it.'

'The evidence suggests,' says Chris French at Goldsmiths College, London, 'that they are not all loonies.'

It is, for sceptics, tempting to put alien experiencers of all kinds in the same file as people who warn crowds of shoppers that the world is ending or who babble incoherently at street corners. It is tempting but, in general, wrong. The truth is that, this one experience aside, these people tend to lead very normal lives and do not exhibit signs of any mental disorder.

This is an important point because it means there is something distinctive about the experience of alien contact or abduction. It is not simply one familiar delusion, interchangeable, in the throes of madness, with any other. Rather it is something that can happen to anyone at any time. If it is real, this is, of course, unsurprising; if it is a delusion, then it is a very odd one. But, whatever it is, the evidence is clear: abductees and contactees are not intrinsically disordered.

Hopkins himself is, indeed, about as disordered as a blueberry muffin. He came into this field, as most do, because he saw a flying saucer.

'In 1964 I had a UFO sighting with my then wife. It was a

daytime sighting over Cape Cod. Up until that moment – I was thirty-three – I had absolutely no interest in or knowledge of the subject, nothing. It came out of nowhere, I sure wasn't looking for it.'

There were three of them in the car. As well as Hopkins and his first wife, Joan, there was an English friend, Ted Rothon, a social worker. They were on their way to a party in Provincetown.

'We were on high ground and we saw to our left, not terribly high up, a small object – lens-shaped and a dull aluminum or pewter colour, not highly reflective. It was just sitting there. There were clouds blowing by, coming in from the ocean. We couldn't really tell if it was moving. I said maybe it's a flat balloon. It didn't seem huge.'

Eventually, by the movement of the clouds, he worked out it was stationary. It couldn't, therefore, be a balloon, which would have been blown by the wind. The car had now reached lower ground and they could look up at the object from beneath. It was circular and entirely featureless. And then it flew off, at, estimates Hopkins, 'about the speed of an airplane'. The whole incident had lasted, perhaps, three minutes.

It was an artists' party in Provincetown. Mark Rothko, among others, was there. Hopkins spoke about the experience. He discovered that several other guests had seen similar things.

'I thought, Jesus, is this going on? It was all new to me. It was not on my horizon at all . . . I have a good relationship with my first wife and I asked her recently, "Can you remember before we saw that thing anybody mentioning UFOs in our house?" And she said no.'

Hopkins is at pains to establish in words what is already evident from his appearance and manner – that he is not a weirdo, UFO-nut or predetermined believer. At the time of the sighting he was not actively seeking confirmation of some faith, conviction or aspiration. It just happened to him, a normal, successful guy with a strong identity and place in the world. The same theme occurs in the stories of Kenneth Arnold and the Hills. It is, I am sure, this normality, this amiable confidence, that has made Hopkins such a key figure in the world of abductees. People trust Budd.

Alien abduction, however, was, at the time of his 1964 sighting, not a respectable topic. There had, of course, always been strange cases of people disappearing, but they had never fallen into any kind of pattern. It frequently, for some reason, happens to soldiers. During the Spanish War of Succession in the eighteenth century, 4000 soldiers along with their equipment and horses vanished. In 1885 600 French colonial troops disappeared, also with their equipment, near Saigon. And, most famously, on 21 August 1915, the 1/4th Norfolk Regiment vanished in its entirety in an 'unusually thick brown cloud wich [sic] seemed to move and rose upward and vanished. There were no traces of the regiment nor their equipment . . .' On 5 December 1945 five US navy TBM-3 Avenger torpedo bombers from Fort Lauderdale vanished with the loss of fourteen crewmen. One ham radio operator is said to have heard Flight 19's last transmission: 'Don't come after me . . . They look like they are from outer space . . . I'm at 2300 feet. Don't come after me.' This may well be the inspiration for the scene in *Close Encounters of the Third Kind* in which World War II planes are found in the desert intact and flyable thirty years after they were lost.

Mystics and occultists had also reported direct contacts with aliens. In 1758 the Swedish mystic Emanuel Swedenborg told of his travels to other planets and his meetings with, among others, Martians. In the nineteenth century the spiritualist Catherine Elise Muller also saw Mars and met its inhabitants. Some spiritualists discovered that when people died on Earth they were taken to Mars. And the extraordinary Helena Petrovna Blavatsky, the leading figure in nineteenth-century spiritualism and founder of the theosophical movement, spoke of the Lords of the Flame on Venus. These Lords were taken up by Guy and Edna Ballard in the 1930s whose doctrines, as Jerome Clark puts it, 'combined popular occultism and native fascism'. Guy Ballard met twelve of the Venusian masters in the Grand Teton Mountains. He created a religion based on contact with extraterrestrials.

But such happenings were isolated eccentricities and, in spite of that radio operator's evidence about Flight 19, lacked an explanatory hypothesis. Furthermore, even after 1947, there seemed to be a reluctance to accept the possibility that aliens might actually appear or do anything other than fly around. The main problem was the number of contact stories that were exposed as hoaxes. George Adamski had obviously been a problem. But also, in 1950, *Variety* columnist Frank Scully published *Behind the Flying Saucers* in which he claimed that in 1947 the US government had recovered three saucers in the southwestern desert. Inside were small humanoid beings from Venus who were 'dressed in the style of 1890'. The link to Roswell was clear. Scully had, in fact, been put on the trail of the saucers by an anonymous lecturer at the University of Denver who said a saucer had crashed '500 miles south of here'. Scully was told he meant Aztec, New Mexico. This seemed credible as, in the nearby town of Farmington, there had just been a significant UFO flap with one witness mentioning a 'huge saucer armada'. Nevertheless, in spite of all these connections, the story was, soon afterwards, revealed as a hoax, though Scully appears to have been a victim rather than the perpetrator. The effect of such incidents was to discredit any reports of contact with aliens. Within ufology, it created the paradoxical state of affairs in which sightings were believed and contacts weren't.

And so, in 1964, at the time of Hopkins' sighting and after seventeen years of widely reported UFO flights, neither actual aliens nor their activities were widely discussed in respectable ufological circles. Whatever Hollywood or the books of the time suggested, the general view was that the UFOs were most likely to be secret military craft or misidentified natural phenomena. Or, even if they were alien craft, people assumed their occupants had made no attempt to contact or interfere with humans.

Then along came the Hills' encounter. Hopkins read the story. His initial reaction was that of the old school ufologists, who desperately wanted to keep their discipline on a respectable, credible scientific basis.

I thought, this just can't be, it's too off the wall. I could accept what I'd seen but I didn't think there was anything inside the object. I was thinking, it's like some sort of camera, a piece of equipment – as if we were on the bottom of a little aquarium and this camera on a long string was being lowered down. But I wrestled with the Hills' story thinking, how can this happen? What's in it for them? The abduction stories were immensely consistent, maybe this really was happening. As David Jacobs, the investigator, said, it took ufologists twenty years to accept that these craft had an inside. We were all so busy trying to get the licence number of the getaway car that nobody had figured out what the crime was.

He mulled this over for eleven years. Then, one day, he fell into conversation with the man who ran the liquor store near his apartment, George O'Barski. They used to talk politics together but, this time, the man was agitated, pacing up and down.

'He said, "A man can be driving home from work and a thing comes out of the sky and little people come out." And he looked at me wondering if I was going to laugh at him. I was thinking, why the hell should he be doing this? He's not a great actor.'

It had happened nearby in New Jersey. Hopkins investigated, found other witnesses to what appeared to be a clear case of an alien landing and subsequently – in November 1975 – published the story in the *Village Voice*. People started contacting Hopkins. Soon he found himself at the centre of the alien contact and abduction phenomenon, gathering together hundreds of accounts. Later he began to use hypnosis to retrieve more details. He currently takes on three or four new cases a week. Cost means that Hopkins has to restrict his investigations to locally accessible places. But the reports come from all over the US and beyond. For, although the vast majority of reported cases are in the US, Hopkins insists that it is a worldwide phenomenon. It is merely less talked about elsewhere and, therefore, less reported.

The most obvious point about the accounts that emerged is the consistency of the story. The basic elements are now familiar from the Hills and from thousands of other similar cases and their use in books and films. Essentially, these are: the UFO, the missing time, the powerlessness of the humans, the control of the aliens over our machinery, usually, as in the Hills' case, a car, the medical examination and/or intervention with its focus on the reproductive organs, the inside of the craft, the imposed amnesia later unlocked by hypnotism and the subsequent physical and mental effects on the abductees. A later refinement involved implants in the bodies of abductees and, among Hopkins' clients, small scoop marks in the skin. The picture emerged of a standard abduction narrative that was to be recounted again and again across the world, as well as being dramatized in TV shows such as *The X-Files* and *Taken*.

Two possibilities arise from the mere fact that such a standard narrative exists at all. First, there is the nuts and bolts explanation: it is really happening and the same stories keep reappearing because the aliens are the same creatures pursuing the same programme. Secondly, there is the psychosocial explanation: it is not really happening, it is some kind of artefact of the abductees' imaginations in combination with the effects of hypnosis, and the reason the same story keeps being told is that people acquire the details, consciously or unconsciously, from their pervasive dissemination through word and image.

Either way, after the Hill case and Hopkins' publication, it rapidly became clear that an extraordinarily large number of abductions were taking place. Eventually, the figures became simply staggering. In 1991 Robert Bigelow, a Las Vegas business-man and founder of the National Institute for Discovery Science, suggested to Hopkins and David Jacobs, another abduction investigator, that a poll should be conducted to find out how many abductees there were in America. It was run by the Roper Organization. Interviewers visited people in their homes and asked them a variety of questions, not all of them to do with abductions. One trick question was put in by Hopkins. People were asked if the word 'trondant' had any particular significance for them. It was an arbitrarily invented word, so if people said it meant something to them they were excluded because their answers would be distorted by an inclination to answer positively.

The obvious problem with such a poll was that, if the aliens had made people forget abduction experiences, then whether a person was an abductee could not necessarily be established by a direct question. Simply asking, 'Have you ever been abducted by aliens?' would produce small proportions of genuine positives –

assuming they had conquered their amnesia – hidden among a larger number of fantasies and an enormous number of negatives. So, instead, the pollsters sought to discover peripheral incidents or experiences that suggested an abduction had taken place. Almost 6,000 people were questioned and the results were startling: 18 per cent had woken paralysed and seen a strange figure in the room, 15 per cent had seen a frightening figure, 14 per cent had had an out-of-body experience, 13 per cent had missing time, 11 per cent had seen a ghost, 10 per cent had flown through the air, 8 per cent had inexplicable scars, 7 per cent had seen a UFO, 5 per cent had dreamt of UFOs and 1 per cent said 'trondant' had some significance for them. A further sifting of the figures by Hopkins and Jacobs indicated that 2 per cent of Americans, some 5 million people, 'have experienced events consistent with those that abductees experienced before they knew they were abductees'.

The figure was bizarre, wild, improbable and the image it evoked surreal in the extreme. To some it was as discreditable to the idea that these abductions were really happening as George Adamski's estimates of Orthon's shoe size were to the idea of contact. If 5 million people had been abducted – many repeatedly, in some cases monthly – in the US alone then the sheer transportation logistics were overwhelming. Showers of people must be constantly hurtling up and down in the skies above America. Furthermore, the lives of repeat abductees ought to be obviously very strange to those who know them. David Jacobs writes, for example, of saleswoman Kay Summer, who lives in the Midwest. She was abducted 100 times in one year.

'Often tired and depressed,' wrote Jacobs, 'Kay has learned to dissociate psychologically from the experience while it is happening, much as a child might during repeated physical or sexual abuse. Still she is on an emotional roller-coaster.'

On one occasion, Summers was abducted while Jacobs was talking to her on the phone.

'She described a roaring noise sometimes associated with the beginning of an abduction, and I could hear this noise over the phone. Hypnosis later revealed that soon after she hung up the phone, she was abducted.'

To deal with the crowds, there must be hundreds of thousands of aliens and hundreds of ships involved. Equally, there must be hundreds of thousands of people who periodically vanish and return without comment. And yet this whole, vast industrial project seemed to have left no trace of physical evidence behind. Admittedly, the absence of individuals can be explained by the fact that the aliens seem to be able to induce amnesia in witnesses as well as in the abductees. But the concealment of the ships would seem to be more problematic. And, in any case, what is the point? If they were studying humans, surely a few, at most, say, a hundred, would be enough.

On the other hand, the pure psychosocial explanation hardly seemed to fit the facts. A large number of otherwise mentally undisturbed people were having extremely strange experiences that seemed to have much in common. The fact that these people were not, as is often thought, exclusively American is evidence against the rapid dissemination of the imagery as an explanation of the consistency of their reports. Hopkins receives reports from all over the world. The former Soviet Union, in particular, seems to have been a hot spot for both UFO sightings and contacts, though it is true that there they did acquire a very local flavour.

On 11 September 1969, for example, Nikolay Zinov was driving his lorry in Kazakhstan. There were three others on board and they got lost. At 5am Zinov stopped the lorry and waited for the sun to rise. He saw a luminous point in the sky. As it approached both he and his companions saw a flying man dressed in a silvery space suit. He landed near the lorry. Zinov approached him and said, 'Comrade, we're lost, show us the right way!' The man did not reply and flew off. Subsequently they discovered they had driven a hundred kilometres past their destination. This implies a possible 'missing time' episode in which they could have been abducted. Had Zinov been hypnotically regressed, he would doubtless have recalled the familiar details.

But, such oddities aside, the information about these experiences – however retrieved – was quite amazingly consistent. It was scarcely adequate to dismiss this vast human airlift as a mass hallucination, a culturally determined artefact or, in Jung's terms, a psychic projection of contemporary anxiety. Moreover, even if the phenomenon could be so defined, then it remained almost as extraordinary as if it were a real, nuts and bolts occurrence. Again, even if it is not real, then it remains most definitely *real*.

As well as Hopkins, there are three other important figures in the history of abduction: John Mack, whose story I shall tell in the next chapter, Whitley Strieber and David M. Jacobs. I separate out Mack partly because his interpretation of these events is quite different – he is, in essence, a third realmer, while the others are nuts and boltists – and partly because his experiences with official disbelief lead into my ensuing theme of alien cover-up. But Strieber is central to the dissemination of the abduction idea.

Whitley Strieber has been a novelist and an advertising executive. Since 1977 he has been a freelance writer. In the eighties, as a result of experiences that began in 1985 when he was forty, he produced the book *Communion: Encounters with the Unknown*.

If only in marketing terms, the most striking thing about this book was its cover. Strieber had asked the artist Ted Jacobs to do a painting of one of the aliens he had seen. A number of key features emerged from the description he gave Jacobs. The mouth – 'Centred in this mouth was a remarkable expression, the outcome, it seemed to me, of implacable will leavened by what I can only describe as mirth.' The chin – 'Very strong, very pointed, and there was a general impression that the skin was stretched over a plated bone structure.' But, above all, there were the eyes – 'They were far larger than our own eyes. In them I once or twice glimpsed a suggestion of black iris and pupil, but it was no more than a suggestion, as if there were optic structures of some kind floating behind those wells of darkness.'

The result of the collaboration between Jacobs and Strieber was an image that was to define the alien face for the next twenty years. The edges of the dark yellow image emerge hazily from a brown background. The focus improves around the features of the face. The chin is pointed and the head rises to a shallow dome. The mouth is thin. So far we could be dealing with a rather odd-looking human. The nose stretches this possibility. It is small and the nostrils seem to be extended downwards. But the eyes – huge, sloping, black and featureless – remove all thoughts that this may be a human face. The image works partly because of its stark geometry but mainly because of its very plausible combination of the almost familiar and the totally alien.

The greatest science fiction often derives its power from the uncanny way the alien intrudes into our world, not, at first, aggressively announcing his presence, but rather insinuating his way in, giving only disturbing and inconclusive hints. What is uncanny about Jacobs' picture is the absolute certainty that this is a face like our own combined with the equal certainty that it is not human. In addition, of course, it is definitely looking at us, but the featureless, black eyes mean that we can have no access to its thoughts or feelings. We are observed but we cannot, effectively, look back.

One investigator, Joe Nyman, told the story of a man and his two sisters who, in 1937 soon after their mother's death, saw a mysterious figure descending the stairs. The man was to see the face of that figure once more, years later, on the cover of *Communion*. Jacobs had painted an image that could not possibly be recognizable and yet, to many, was.

'The large head with the now characteristic staring alien eyes,' wrote David Jacobs, 'considered by most abductees to be anatomically incorrect, triggered memories that impelled thousands of people who thought they might be abductees to seek advice and help from abduction researchers.'

Those eyes alone must have sold thousands of copies. Certainly they were instrumental in making this perhaps the most influential text of abduction lore. Strieber became, as Jacobs put it, 'a media icon for the abduction phenomenon'. He is an accomplished writer and his constant self-questioning about his experiences dramatizes the feeling that this is a normal man under abnormal stress rather than a credulous fanatic.

'To all appearances I have had an elaborate personal encounter with intelligent, nonhuman beings . . .' Strieber explains. 'Later I found a large number of people who have had experiences similar to mine. Most of them were mentally stable.

They did not cluster in a particular population group, but formed a cross-section of American society.'

In the middle of the night on 26 December 1985 in a cabin in upstate New York, Strieber is woken by a 'peculiar whooshing, swirling noise coming from the living room downstairs'. He tries to get back to sleep, but then he sees a small figure edging into his bedroom. The figure rushes at him and then there is only blackness. He is aware of movement but little else. Then more aliens seem to be involved. Ultimately he sees four different types: small robot-like creatures, short, stocky ones in dark blue overalls, a group of slender, delicate ones with slanting black eyes and small, huddled figures with black eyes 'like large buttons'.

One alien opens a box containing a long, thin needle which it plans to insert into Strieber's brain. He imagines his family waking in the morning to find him a vegetable – 'You'll ruin a beautiful mind,' he protests. For reasons he cannot explain, he asks the aliens if he can smell them. One alien acquiesces. Its overall has a slight scent of cardboard, its hand a certain sourness.

'It was not a human smell, but it was unmistakably the smell of something alive. There was a subtle overtone that seemed a little like cinnamon.'

Then: 'The next thing I knew I was being shown an enormous and extremely ugly object, gray and scaly, with a sort of network of wires at the end. It was at least a foot long, narrow and triangular in structure. They inserted this thing into my rectum. It seemed to swarm into me as if it had a life of its own. Apparently its purpose was to take samples, possibly of fecal matter, but at the time I had the impression that I was being raped, and for the first time I felt anger.'

Subsequently, Strieber contacts Budd Hopkins – 'a large, intense man with one of the kindest faces I have ever seen' – who listens sympathetically to his story and elicits an earlier experience of what seems to have been a visitation at the cabin. Hopkins wants him to see a therapist, Dr Aphrodite Clamar, but Strieber insists on seeing somebody who has had no previous involvement with alien abduction cases.

He visits Dr Donald Klein of the New York State Psychiatric

Institute. Dr Klein was to conclude that Strieber 'is not suffering from psychosis' and was not hallucinating 'in a manner characteristic of psychosis'. Under Klein's hypnosis, he sees more details of his experiences.

'Looking down at me. Got eyes. Big eyes. Big slanted eyes. A bald head. He's looking down at me. He's got a ruler in his hand. Has a tip of silver. Touches me. I see pictures. (*Long pause*) I see pictures of the world just blowing up. I see pictures of the whole place just blowing up when he touches my head with this thing.'

This seems to accord with Strieber's own conviction that the world is 'in pretty serious trouble' and that 'maybe the idea of visitors coming along and saving our necks was more appealing to me than I might consciously have wished to admit'. In *Communion*, Strieber remains confused to the end about what exactly is going on. But he inclines strongly towards the nuts and bolts benign alien hypothesis. Indeed, the consoling title of the book was given to him by his wife, talking in her sleep in an alien voice. Apparently she was being used as a conduit to communicate to Strieber.

'The book must not frighten people,' said the voice, 'you should call it "Communion", because that's what it's about.'

Like many others, Strieber suspects that the aliens are here because of our spiritual failings.

'It would seem that our civilization is not paying attention to what may be the central archetypal and mythological experience of the age. If so, then this is the first time that man has simply refused to respond to the ghosts and the gods. Is that why they have become so physical, so real, dragging people out of bed like rapists in the night – because they *must* have our notice in order to somehow be confirmed in their own truth?'

Strieber's book is, perhaps, the key document of the abduction phenomenon. The cover, the lucidity of the writing, the normality of the author, the precision of his observations and his genuinely tentative speculations about meaning all combined to make it both vivid and respectable. It could persuade or at least intrigue sensible doubters in ways that much of the more florid abduction literature that had begun to emerge could not.

Communion prepared the way for the TV series *The X-Files*,

which began in 1993. Created by Chris Carter, this was predicated on the idea that there was, indeed, a massive alien
operation taking place around us. Much of the imagery seemed
to derive from Strieber. However, the X-Files operation is
definitely malign and it is being pursued in partnership with
subversive elements within government. Presumably to keep the
series running, the answer to what exactly is going on is never
quite given and, as in Strieber, we are left with an infinity of
speculation driven only by the show's motto 'The truth is out
there'.

This uncertainty, however, is not entirely a convenient struc-
tural matter for television producers; it seems also to be built into
the contact and abduction experiences. Thomas E. Bullard has
observed that abductees 'feel a sense of mission but they seldom
know what their mission is'. Something important has plainly
happened and abductees feel they have been singled out. They
must, therefore, do something. But what? Should they resist the
aliens and warn of their coming or should they welcome them
and help advance their project on Earth?

On the whole, however, abductions are interpreted as malign
for obvious reasons. Strieber's benign interpretation is remarkable in view of what appears to have happened to him – the
needle in the head and the gigantic anal probe. Most abduction
experiences seem to be very traumatic, hence the need for
insurance. Many, if not most, investigators have, as a result,
concluded that the aliens definitely do not mean well.

'Shall we say I don't take an optimistic view?' says Budd
Hopkins,

> . . . I tend to be conservative in the sense that I don't know
> what ultimately is to come of this, but I've just seen too
> much psychological and emotional damage among
> abductees. These aliens are not considerate. They're not
> brutal, exactly, it's just that they do what they want and
> there's no emotional resonance . . .
>
> I think, whatever they're doing, is for them. I think, to be
> very specific about it, that they've reached some kind of
> evolutionary dead end. I think they need somehow to revive

their species. Humans are very primitive beings by their standards, but we have some kind of vitality that they lack . . .

This may sound very strange, but, at this point in my life with a 30-year-old daughter who wants to have a family . . . I think that if and when this happens, the world is going to be changed radically and it's going to be so radical in one way or another that, if you happen to like the world the way it is, as I do . . . Well, I don't look forward to that.

The most effective advocate for the malign hypothesis is David M. Jacobs, a professor of history at Temple University in Philadelphia. In his book *The Threat: Revealing the Secret Alien Agenda*, Jacobs rejects all efforts to portray the alien project as an attempt to save humanity.

A careful examination of the UFO abduction phenomenon shows us that contact has, in fact, occurred – but it bears no relationship to these scenarios. There has been no public meeting, no involvement of leadership, no press coverage. There has, as yet, been no assistance, no cooperation, no death and no apocalypse. The contact has been on the aliens' terms – and secret . . .

Virtually everything that aliens do is in service to their abduction program. Every seemingly incomprehensible or absurd alien activity has, upon examination, a logical basis. One by one, these actions have begun to lose their mystery and reveal their true purpose.

Jacobs believes the persistent emphasis on human environmental degradation of the planet may be a diversion as they 'almost never say or do anything to help the environment; they only lament its desecration'.

Their true purpose is the 'Breeding Program' 'in which the aliens collect human sperm and eggs, incubate fetuses in human hosts to produce alien-human hybrids, and cause humans to mentally and physically interact with these hybrids for the purposes of their development'. This programme started early in the twentieth century and may have produced hundreds of thousands, maybe millions, of children.

This would, of course, explain the sexual theme that comes up in so many abduction accounts. From Antonio Villas Boas onwards, alien interest has seemed to focus on our reproductive systems. Hopkins has noted that, though human doctors tend to focus first on our cardiac, vascular and respiratory systems, aliens tend to go straight for the genitals. Alien sex is, of course, both, thrilling and chilling. Ian Watson's novel *Miracle Visitors* is a clever fictional accumulation of real alien lore and includes this account of a sexual encounter between a human man and an alien woman, complete with the Villas Boas dog sounds.

'She pulls me into her, on the couch. And it hurts me! Oh, I'm enjoying it, but there's a cold pain in my balls as if they've been dipped in a bucket of ice. I never knew it hurt a man the first time! Now's the only time she makes any sound: like a softly growling dog, a growling bitch –'

The idea of a man feeling pain during his first sexual experience shows a subtle awareness of the implications of sexual abduction. For, of course, it is women who usually feel pain the first time. But a man is partially emasculated if he is abducted and forced to have sex or, as often happens, to donate semen. He is the passive partner. As a result, the experience is frequently embarrassing – Barney Hills, remember, was too ashamed to report the extraction of his semen until long after the incident. For women, the experience is equivalent to rape and may also, therefore, be suppressed as too appalling or embarrassing.

But the idea of alien sex remains enticing as it offers an escape from the vicissitudes of human sex. This was an important theme in the sixties and seventies when the first abduction cases were appearing, because human sex was suddenly felt to have been distorted by political and social oppression.

The sexual revolution of the sixties represented a rejection of the suburban mores of the fifties. These mores were based on that decade's unique combination of anxiety and affluence. The tight-knit, sexually faithful couple sustained the affluence and kept anxiety at bay. Chaste sex kept the cosy suburbs intact. Imaginatively, it protected affluent society from the threats that lay beyond the suburbs: nuclear war and environmental decay, of course, but also social disintegration.

In the sixties, this protection began to unravel. Sexually pure consumers in their perfect houses became objects of derision. The underlying conviction of the revolutionaries was that the suburb was the location of the most fearsome bourgeois sexual oppression. Feminism added the conviction that men were the oppressors. They did not want real women, they wanted sex slaves and passive housewives.

This was the thesis of perhaps the most celebrated alien, in the sense of robotic, sex film of all, *The Stepford Wives* (Bryan Forbes, 1974). The men of the town have gradually replaced their wives with sexually, emotionally and practically compliant robot simulations. All their troublesome features – moodiness, independence of thought, fatigue – have been removed. This, it is assumed, is what men really want: female machines in flowery dresses and big hats, fantasy-generated aliens. The town has become a conspiracy of the male bourgeois against the female. Similarly, in *Westworld* (Michael Crichton, 1973) the bored bourgeoisie are shown escaping to a series of theme parks – 'the vacation of the future today!' – where they can harmlessly live out their fantasies amidst a cast of robots. Nobody gets hurt and, each night, the robots are repaired by the parks' technicians. Sex is necessarily one of the pleasures on offer. In the Wild West setting of his theme park, the hero indulges himself with a local whore, but, unseen by him, her eyes simply open in a blank stare at the climax of the act. 'I think you're very nice,' she says as she leaves. The moment is chilling because it indicates one real male anxiety of alien sex – that your excitement is controlled by rather than shared by another, that you are doing little more than masturbating to order. Aliens always fake orgasms.

The seventies, in fact, was a decade riven with uncertainty about human sexuality. Also in 1973, Ursula K. LeGuin published her novel *The Left Hand of Darkness*. There is a planet named Winter or Gethen. An envoy arrives here from Ekumen, a group of planets inhabited by humanoids who are all thought to originate from one planet, Hain. The inhabitants of Gethen are asexual, but, once a month, they enter Kemmer in which they change, more or less at random, into males or females. A sexual fluidity is implied, consistent with the emerging belief of the time that gender was no longer to be seen as an absolute. Anybody can bear a child and the child has no psychosexual relationship to its father or mother so there is no Oedipal conflict. There is no non-consensual sex, no rape, no division of people into strong and weak, active and passive. LeGuin had simply projected the possibility of a utopian sexual fluidity on to an alien species.

Both movies and the novel express a very seventies feeling that sex had become an arguable issue rather than a dark absolute as well as the fear that the revolutionaries of the sixties might be imposing a mechanization and desacralization on human relations. Instead of an encounter between persons, sex becomes an encounter between person and machine. The men, particularly in *The Stepford Wives*, are revealed as the desacralizers because they are seen as content to have – indeed, they prefer to have – sex with robots. But, in fact, both sexes are complicit in this process. It was Erica Jong who, in *Fear of Flying* in 1973, coined the phrase 'the zipless fuck' meaning a sex act purely for its own sake, devoid of wider significance.

'The zipless fuck,' she wrote, 'is absolutely pure. It is free of ulterior motives. There is no power game. The man is not "taking" and the woman is not "giving" . . . the zipless fuck is the purest thing there is. And it is rarer than the unicorn.'

But the zipless fuck is also desacralized sex and, in this context, Jacobs' Breeding Program can be seen as a nightmarish vision of the logical outcome of such encounters. The alien fuck is the supremely zipless fuck in that it is even more completely devoid of context. A woman knows what a man wants from the encounter, but what do these guys want? Our sexuality is simply being used for non-human ends. But what are these ends?

The answer appears to be that the aliens are in the midst of some form of genetic crisis which means they have to interbreed with humans in order to revivify their species. They suffer, in some accounts, from an excessively narrow gene pool and, as with threatened animal species on Earth, this results in a dangerous enfeeblement. The variety of our species – and this is repeated many times by many different aliens – is unique in the cosmos. It is our greatest strength and our greatest attraction to species less various.

That said, abduction narratives have revealed that the aliens do, at least, have a more radical form of variety in that they come in several types. There are taller, 'Nordic' type aliens. They generally seem to be in charge. The smaller Greys seem to be the worker aliens. Some suggest they are robot replicants. There are other reptilian and insect types – one, a 'praying mantis type' is specifically said to need human genes to reproduce. But, one way or another, they all seem to need us for some reproductive assistance.

Furthermore, we have an emotional and spiritual depth they lack. This is, note, the exact opposite of the view that aliens regard us as spiritually deprived. That belongs to the benign interpretation of alien presence. In the malign interpretation, we are spiritually gifted and they wish to employ this gift for their own purposes. One of Jacobs' abductees, Reshma Kamal, is asked by a female alien to hold a hybrid baby. She resists and demands an explanation.

> She thinks if she can make me understand something, I'll behave better . . . I know she's trying for me to cooperate. I'm thinking the more I bother her, the more she will give me information. Now she's telling me that they need these babies. What we need to teach them is emotions, feelings, that they cannot do. She's explaining to me that they can feed and clothe the babies, they can grow physically, but they cannot give these babies emotional development, that they need me to help them do that. I don't understand that . . . She's saying there's a very big need for these babies. She's saying something about these babies are not exactly like them, or not exactly like us. But they need to have emotion . . .

The aliens want to use our souls, the 'magic touch' we apparently possess, to enhance the development of their hybrid children.

The overall programme, according to Jacobs, has four stages:

abduction, breeding, hybridization and, finally, integration, at which point hybrids or aliens themselves are inserted into human society and, ultimately, take over. What we are really seeing in these mass abductions, therefore, is a prelude to an invasion. This prelude is marked by surgical and sexual desecration. This interpretation also explains the scale of the abductions. Plainly, millions of humans are needed for what is, in effect, the creation of an invasion force.

There are, however, two further aggressive forms of intervention by the aliens: implants and cattle mutilations. There is also the seemingly more benign phenomenon of crop circles. These do not fit quite so neatly into this theory, though neither do they refute it. Different aliens might be involved, or these could simply be further aspects of the invasion programme.

Implants are familiar from *The X-Files* and countless other SF shows and films. Abductees have small objects placed inside them which subsequently show up on X-rays, cause bleeding or spontaneously emerge. Frequently they are elusive, disappearing when investigated and, even when found, their form and function remain inscrutable.

However, Dr Roger K. Leir, a podiatrist from Southern California, seems to have had more success than most with implants. In his book *The Aliens and the Scalpel: Scientific Proof of Extraterrestrial Implants in Humans* he shows that he has, indeed, acquired some of these implants. In a foreword to the book, Whitley Strieber says that children in the future will learn these crucial dates: 11 June 1985 – Leir hears about implants at a lecture in Houston; 19 August 1995 – Leir removes three objects from the feet and hands of abductees' 18 May 1996 – further surgery with witnesses attending, including Strieber, who describes the occasion as 'among the most moving experiences of my life'. Strieber himself has twice been implanted – on one occasion deep in the middle of the left temporal lobe of the brain, far beyond the reach of surgery but visible to an MRI scan.

Leir had been interested in ufology since the Roswell incident in July 1947. He recalls his father walking into the kitchen and announcing that the USAF had captured a flying saucer. He held up the newspaper headline and went on to explain his conviction that the aliens had, indeed, arrived and that the 'powers that be' were keeping it secret. Leir has also had two NDEs – near-death experiences – once as a result of inhaling an excess of nitrous oxide, thinking it was oxygen, and once as a result of a plane crash. On both occasions he felt himself entering a peaceful realm but, each time, a stern voice commanded him to 'Go back!'

Leir approached the idea of retrieving implants with some trepidation. Previous attempts, even when successful, had been inconclusive. Richard Price, for example, had removed an implant from his own penis and passed it on to Dr David Pritchard at Massachusetts Institute of Technology. Electron microscopy failed to find any evidence of extraterrestrial origin. 'The final conclusion,' notes Leir, 'was that the object was interesting and remained an unknown.'

Leir himself claims to have removed a number of implants and noted, for the first time, anomalous tissue behaviour in the vicinity – notably a lack of inflammation and an unusual number of nerve proprioceptors. But, most importantly, 'most of the metallurgical analysis indicates that the structure of these objects has an extraterrestrial origin. . . Moreover, the form of these objects is clearly engineered and manufactured with precision rather than being a naturally occurring form.'

'If,' Leir concludes, 'these scientifically derived results are not disproven by subsequent analysis, then we can firmly conclude that *some individuals with alien abduction histories have artificially manufactured objects in their bodies or a demonstrably extraterrestrial origin.* In short, we now have the 'smoking gun' of ufology – hard, physical, scientific evidence of a continuing alien presence on Earth!'

The alien implant is, plainly, thematically linked to alien sex. Both involve an invasion of the body that seems more frightening than the mere invasion of the Earth that UFOs had previously threatened, and both, as Leir says, imply a continuity, an ongoing project in which we, whether we like it or not, are involved. This development represents a progression inward. If aliens were associated with external, public threats and anxieties in the forties and fifties, in the sixties and seventies they came to be linked to private, intimate threats. Nuclear fallout represents one form of anxiety; an anal probe another.

There is a further link – and it is not a tenuous one – to a change that happened in the course of the same period in science. The structure of the DNA molecule was deciphered in 1953 by Watson and Crick. At the time it had nothing like the impact of the advent of the nuclear age. But it marked the moment at which physics began to give way to biology as the most promising or threatening science of our era. Physics had offered to change the world; biology was to offer to change ourselves. A decade or more after Watson and Crick's achievement, this began to dawn on the popular imagination. Now, of course, it is a common-place. The effect was to redirect technological and scientific anxieties inwards. The idea of the alien within became much more scary than the blob, the giant spider or the vegetable man without.

'Abduction experiences,' writes Erik Davis in his book *TechGnosis*, 'partly speak to the subconscious horror induced by the reduction of human identity to a twisted strip of genetic information that can be spliced and diced like a filmstrip.'

DNA implies that we are 'just' our genetic code. The alien within, therefore, can simply rewrite it as we would a computer program.

Greg Bear's novel *Blood Music* (1985) most precisely drew together the themes of genetics and the alien within. Vergil Ulam, a renegade biologist, injects himself with a line of cells in order to smuggle them out of a laboratory from which he has been sacked. The problem is that he has been working on making these cells intelligent and, once they are inside him, he discovers that he has been successful. These are truly alien beings.

'Are they friendly?' asks his mother.

'They're lymphocytes, Mother. They don't even live in the same world we do. They can't be friendly or unfriendly in the way we mean the words. Everything's chemicals for them.'

They begin to improve Vergil's body – 'See? I'm being rebuilt from the inside out.' And why are they doing this?

'So they don't have to rely on us anymore. The ultimate selfish gene. All this time I think the DNA was just leading up to what

I've done. You know. Emergence. Coming out party. Tempting somebody, anybody, into giving it what it wanted.'

DNA itself has become the invasion force and it is far more alien than any aliens we have previously encountered. To ask, as Vergil's mother does, whether it is friendly or not is beside the point. There are no such categories in the world of bacteria or lymphocytes – or, as they become once they acquire intelligence, noocytes.

With contacts, abductions, surgical and sexual interventions, implants and internal invasions, the field of what had once been called ufology had expanded to take in many neighbouring realms. But, though it had expanded in its range of expression, it had contracted in its physical area of operations. It was now inside the human body. But there was one new form of alien intervention that seemed – but only seemed – to take us back into the outside world. This is the bizarre phenomenon of cattle mutilations.

The unexplained mutilation of livestock – removal of genitals, draining of blood, missing ears, tongues, eyes or udders, flayed patches and 'cored out' rectums – has produced more hilarity among sceptics than any other alien manifestation with the possible exception of the anal probe. Even the books on the subject can't seem to fight off the comic possibilities. There's *Altered Steaks* by David Perkins and Tom Adams and *The Choppers – and the Choppers: Mystery Helicopters and Animal Mutilations* by Tom Adams. In the *Toronto Star* of 17 December 1979 Corporal Lyn Lauber of the Royal Canadian Mounted Police was quoted as saying, 'I can't see what the attraction of a bull's ass would be to a UFO.' But this kind of comic scepticism cuts, as it were, both ways. For Jacques Vallée, for example, cattle mutilation seems to be an argument against any psychosocial explanation. 'If spacemen are thought images,' he asked, 'why should they butcher a cow and drop her head in the desert?'

Butchered cattle seem to be real, nuts and bolts evidence. And yet, at the same time, it is evidence so bizarre and crude that it contrasts violently with the apparent delicacy and refinement of alien technologies seen by abductees.

Aliens had, from the beginning, taken a special interest in

livestock. On 16 October 1954 a farmer in France saw his horse drawn up into the air and a hunter in the US had seen the same thing happen to an elk. But the first case of mutilation seems to have been on 5 October 1969 when a three-year-old horse was found dead on a ranch in Colorado. Its head had been completely stripped of flesh and muscle. There were no signs of a predator and there was a strange medicinal smell in the air.

Further cases began to emerge in Wisconsin, Kansas, Oklahoma and Minnesota in the early 1970s, within a few years of the first wave of abduction stories. During the course of that decade mutilation cases were to run at 500 a year. The first mutilations were assumed to be perpetrated by humans. But the numbers were prodigious. David Perkins, co-author of *Altered Steaks*, says that by the end of 1975 'phantom surgeons had struck in virtually every state west of the Mississippi'. Furthermore, there was an absence of blood and the cuts were unbelievably neat. They could have been done by lasers, but surgical lasers were not then available.

Whoever was doing it, losses to farmers were running into millions and Governor Lamm of Colorado called it 'one of the greatest outrages in the history of the Western cattle industry'. The crisis led to a multi-state mutilation conference in Albuquerque in April 1979. At this event, Tom Adams, a researcher from Texas, revealed the link between mutilations and 'black helicopters' – there had been 200 cases where these craft had been seen in the vicinity of a mutilation. The implication was that there was some kind of government–alien hook-up.

Like everything else in the realm of aliens, cattle mutilations led to a bitter battle between sceptics and believers. On the one hand there was the deflating conclusion of a report by ex-FBI man Kenneth Rommel, appointed by the Albuquerque conference as project director of Operation Animal Mutilation. He simply said all the fifteen cases he investigated were the work of natural predators. On the other hand, there was the incredible onslaught in Cascade County, Montana. In nine months there were 100 incidents and 130 reports of mysterious helicopters or UFOs. These incidents took place in the vicinity of Malmstrom Air Force Base and close to a number of Minuteman missile silos.

What Perkins calls 'high strangeness' began to emerge in Cascade County when two women reported meeting a seven-foot tall creature with a 'dark and awful' face and a man encountered a tall hairy creature carrying an object the size of a bale of hay.

The mutilations could be fitted into the Jacobs Breeding Program hypothesis. It was suggested, for example, that the animals' enzymes or hormonal secretions were essential to the survival of the troubled aliens.

'We use substances from cows,' one alien message explained, 'in an essential biochemical process for our survival. The material we use from cattle contains the correct amount of protein substances needed for biochemical absorption . . . While we respect all life, some sacrifices must be made.'

One abductee, Judy Doraty, recalled, under hypnosis, seeing a calf being drawn up into a UFO by a beam of light. Once on board the ship herself, she saw its organs being surgically excised. The aliens told her they were testing our soil for poisons and carrying out tests on animal reproductive systems.

But, perhaps because of the high strangeness of the mutilation phenomenon, many theories were much more radical and specific than Jacobs'. In the minds of some, this all pointed in one direction – to the Dark Side. The Dark Side hypothesis was a combination of the ideas of Bill Cooper, an ex-naval petty officer, and John Lear, a former airline pilot who appeared to have CIA connections. This hypothesis tied together abductions and mutilations into a grand paranoid synthesis. Lear claimed to have proof of an underground base near Dulce, New Mexico, where aliens and the US government worked together. Here there were hybrid embryos that used genetic material from both cattle and abductees. Lear believed that there had been dozens of saucer crashes and the government launched Operation Redlight in 1962 with the aim of discovering how to fly these machines. An area of land near Area 51 has been handed over to the aliens who are allowed to abduct as many people as they want. In return they offer their technology to the government. A dispute arose in the seventies which resulted in the aliens killing forty-four scientists and some Delta Force troops sent in to rescue them. The alliance was resealed in the 1980s and the abduction programme was continued with the approval of a secret US government body known as Majestic 12 (MJ-12), the committee believed to have been in existence since the forties. Cooper, however, insisted the 'M' stood for Majority and added a few refinements of his own. I cannot improve on this summary of Cooper's revelations from www.worldofthestrange.com.

All the while flying saucers were dropping like flies out of the heavens. In 1953 there were 10 crashes in the United States alone. Also that year, astronomers observed huge spaceships heading toward the earth and in time entering into orbit around the equator. Project Plato was established to effect communication with these new aliens. One of the ships landed and a face-to-face meeting took place, and plans for diplomatic relations were laid. Meanwhile a race of human-looking aliens warned the U.S. government that the new visitors were not to be trusted and that if the government got rid of its nuclear weapons, the human aliens would help us in our spiritual development, which would keep us from destroying ourselves through wars and environmental pollution. The government rejected these overtures.

The big-nosed grays, the ones who had been orbiting the equator, landed again, this time at Holloman AFB, in 1954 and reached an agreement with the U.S. government. These beings stated that they were from a dying planet that orbits Betelgeuse. At some point in the not too distant future, they said, they would have to leave there for good. A second meeting took place not long afterwards at Edwards AFB in California. This time President Eisenhower was there to sign a formal treaty and to meet the first alien ambassador, 'His Omnipotent Highness Krlll,' pronounced Krill. He, in common with his fellow space travelers, wore a trilateral insignia on his uniform; the same design appears on all Betelgeuse spacecraft.

The X-Files was, in effect, a fictional version of this Dark Side hypothesis.

In contrast to all this darkness, there are crop circles, a phenomenon centred in Britain, though, once again, said to occur all over the world. Also they are said to have been appearing for centuries, though they only came to wide public attention in the 1980s. Many have been revealed as hoaxes, though a widespread faith in their authenticity remains. These are said to be alien signs. They may, indeed, be malign, but, on the whole, they are taken to be messages sent from a benevolent cosmos, pastoral but indecipherable greetings from the heavens.

They are, however, an exception to the increasingly dark body of evidence that has dominated alien lore since Betty and Barney Hill's abduction was publicized. Perhaps darkest of all is the way humans seem to have exacted very precise revenge on the alien surgeons. In 1994 Ray Santilli, a London-based film collector, acquired some old black and white footage. It showed autopsies being conducted on the alien corpses recovered at Roswell in 1947. In March 1995 the story broke via the Press Association: 'A top-secret film allegedly showing dead aliens will be screened for the first time anywhere in the world this summer – in Britain.' The argument about the authenticity or otherwise of this footage continues and, indeed, more film has since emerged.

But, true or not, the black humour of the situation is inescapable. The aliens arrive. We cut them up and they cut us up. We haven't yet laid our hands on any of their livestock, but doubtless we'd core out their rectums if we did. The alien encounter, it seems, is nothing if not surgical.

5

John Mack

In June 1994 the Committee for the Scientific Investigation of Claims of the Paranormal (CSICOP) staged a conference in Seattle entitled The Psychology of Belief. John Mack, professor of psychiatry at Harvard, was there. He had just published his book *Abduction: Human Encounters with Aliens* and he had become, as a result, a media event, appearing on the Oprah Winfrey and Larry King TV shows. The big news was that a Harvard professor had said the aliens were real.

Mack spoke, declaring abductions were 'authentic mysteries'. He said the phenomenon was inviting us to stretch our understanding of reality. 'Other cultures have always known that there were other realities,' he said, 'other beings, other dimensions. There is a world of other dimensions, of other realities that can cross over into our own world.'

Mack also wondered why the sceptics were so vehement in their attack on abduction claims. 'Are we,' he asked, 'seeking to be the arbiters of reality?'

Later a researcher called Donna Bassett rose to speak. At first she gave the impression of being one of Mack's abductees and she had, indeed, participated in his work. But then she said she had, since September 1992, simply been posing in order to infiltrate Mack's operation.

'I faked it,' she said. 'Women have been doing it for centuries.'

Mack's procedures, she claimed, were flawed and, as a result, the abductees just 'told him what he wanted to hear'.

Mack expressed disappointment at Bassett's speech. 'I'm not convinced one way or the other, whether she did in fact hoax or whether she has in fact had these experiences herself. I don't know.'

Mack also said she had been put up to this gesture by Philip J. Klass, one of the most aggressive and unstoppable sceptics in the UFO business. Klass, who was in the audience, lost his temper, stormed up to the stage and accused Mack of making 'false innuendos'. The row went on until the chairman finally brought it to a halt.

Mack's involvement with abductions was to lead him into yet more trouble. The Harvard authorities became concerned about his belief in the reality of these alien contacts. An inquiry was launched to look into his therapeutic procedures. But the issue rapidly became one of academic freedom. Eventually Harvard climbed down, Mack remained there as professor of psychiatry and continued to publish books based on the hundreds of interviews he had conducted with abductees.

He is a crucial figure for two reasons. First, his story embodies the theme of a cover-up. As I shall outline in the next chapter, this idea is at the centre of alien lore. Usually it is a government cover-up, but, in Mack's case, it was academic. Secondly, Mack's interpretation of the abduction phenomenon was quite different from that of Jacobs, Hopkins or Strieber. He was a third realmer, an interpretation covered in more detail in chapter 8. He believed the alien presence is benign and offers the possibility of the rescue of the Western mind and soul from the crass materialism into which they have flung themselves.

Mack was an intense man who talked slowly and quietly. He seemed to be constantly on the alert for hidden significance in the world. He was, for example, interested in the coincidence that I had contacted his office just as he was about to come to London. During our talk and afterwards, when I took him to Paddington

Station, I had the impression of an uneasy man, somehow uncomfortable with things as they appeared to be. An attempt to leave his luggage at the check-in desks at the station failed. I had known it would, not because I knew anything of the system, but because I could see the intense and preoccupied Mack was not on easy terms with the world. (An email from the Psychology of the Paranormal Network informed me of his death in September 2004. He had been hit by a car in North London. The banality of the circumstances clashed painfully with his exotic and ambitious preoccupations and I remembered my strong sense of a man poorly attuned to the vicissitudes of life. I should have liked to see him again.)

He was brought up in New York City by German-Jewish intellectual parents. It was, he pointed out, a double-edged cultural background – on the one hand highly scientistic, 'very left brain', on the other hand a culture of great bravery and curiosity. His upbringing was conventionally academic but not scientific. Indeed, he never thought of himself as a scientist.

'Most of the science courses I took didn't grab me and I think I now understand why. I guess now I would see that material science is divorced from something more fundamental. It's purely facts about the physical world which are interesting but they're not the whole story. This is still an evolving realization.'

He wrote a biography of T. E. Lawrence, *A Prince of Our Disorder*, but, primarily, he was a psychiatrist, a leading figure in Harvard Medical School. His pursuit of his subject, however, was always unconventional. He became very involved with attempts to stop the arms race in the seventies – 'a political system that would risk the incineration of the world'. And he was a leading figure in International Physicians for the Prevention of Nuclear War which, as a body, won the Nobel Peace Prize in 1985. This led him into a project in which he interviewed twenty people involved in nuclear decision making, including Jimmy Carter, about their motivations and justifications. He found it was 'a whole boys' club . . . you couldn't question what you were doing, that would be self-emasculating'.

He was also to become involved with Erhard Seminars Training or EST. This was a movement started by Werner

Erhard, formerly John Rosenberg, which consists mainly of what Mack called 'a technology for blowing your mind, basically'. EST, with its highly disciplined and exotic structures of meditation and training, was many things to many people, but, as far as Mack was concerned, it made him begin to 'question everything'. He also took up Stanislav Grof's holotropic breathwork; this involves using rapid breathing to enter an altered state of consciousness.

'I found that it opened my consciousness to a whole other way of thinking about the psyche . . . I travelled into past lives, emotions and events. I realized the psyche could travel. It was not limited to the brain and the body. Spirituality, rather than being an embarrassing high-mindedness, which is what it is in secular culture, became very tangible.'

It was, he admitted, a big step from that to alien abduction, but 'the world beyond the material world had begun to take on a palpable reality'.

His interests had strayed far from conventional psychiatry and yet he remained one of the leading lights of Harvard Medical School. Then, in 1989, he was given an article on UFOs. He began to wonder if this was true. In 1990 he met Budd Hopkins and heard of his vast range of abduction accounts.

'I realized immediately that this was the way people talked about things that had happened to them. They didn't sound like fantasies.'

He arranged to meet some of the abductees.

'They struck me as ordinary people who had had extraordinary experiences. They were very distressed about what had happened. . . If you went outside the realm of possibility in the Western ontological, ideological tradition, there would have been no doubt in your mind that these people had actually experienced what they said.'

Mack's words here are crucial. He is raising the question: why should we disbelieve these people? What is it in *us* that makes us doubt *them*? We wouldn't be inclined to think them liars or deluded if they said they had gone shopping or been on holiday, if, in other words, they stayed within a certain frame of reference. But that frame of reference is a remarkably narrow, anthropological oddity. Outside our Western, scientific, secular culture, people would be unamazed to be told somebody had been communicating with spirits or ghosts. It is normal. The default human belief condition is that there is another world in close proximity to ours and the two routinely interact. We are the aberration.

> Throughout most of the last three hundred years, Western thought has been able to tolerate the notion that we can reach the spirit world. But the notion that the spirit world can cross over and show up here in some tangible form is absolutely not within the realms of ontological possibility. That's the key thing, the crossover phenomenon . . .
> I was faced with a choice. Did I somehow try to do what other people have done, which was to squeeze this thing into a structure of thought? Or did I decide there was something limiting in that structure of thought? The second choice seemed more intellectually honest.

For Mack, the deal made in the seventeenth century that separated the material world of science from the spiritual world of religion began to break down. It was always, in his terms, an impossible division. There are perpetual interactions between the two realms and only our secular culture denies this obvious truth. The Donna Bassett incident was, for Mack, an example of the lengths to which our culture will go to protect its world view.

'Barry Goldwater once said that extremism in the defence of liberty was no vice. Well, it seems that extremism in the defence of Western scientific materialism is no vice either.'

Mack was gripped by the ferocity of the defence of the West. He wrote in his book *Passport to the Cosmos*: 'To destroy someone's world view is virtually to destroy that person.' And he spoke of the Zen concept of *daishi-gyo*, the great death that 'occurs when there is a melting away of our images of self, the world and culture – the loss of the sense of who we are in the cosmos that a world view provides'. As a result, the more irate and intemperate his critics, the more convinced he became that something fundamental in their lives was being challenged by what he said. Hostility didn't kill him, it made him stronger.

In reality, 'the most important truths for our culture may lie in the extraordinary nature and power of the abductees' experiences, the opening that these experiences provide to other deeper dimensions of reality, and what they may mean for our culture and the human future'.

Mack's work with abductees convinced him that the materiality of the experiences was not the issue and that the use of a purely empirical approach to a search for evidence would not succeed. That is to play the scientist game. He was not, in other words, a nuts and boltist. Alien abductions are, like near-death experiences, crop circles, apparitions, inexplicable healing powers and parapsychology, phenomena that 'are forcing us to appreciate that cosmic realities exist beyond the three-dimensional universe that has bounded our earthly existence'. If we open our consciousness we will discover a cosmos 'filled with beings, creatures, spirits, gods . . . that have through the millennia been intimately involved with human existence'.

What appears to be happening now is that we are experiencing some kind of feedback, a response from the spirit world. This is intended to stop our self-destructive ways, particularly in relation to the environment. But the aliens are not simply here to save us from ourselves. They need something in return – embodiment.

'Some abductees are informed that something went wrong biologically for the aliens as a result of an over-stepping or technological hubris of the kind we are engaged in now on this planet, and like the angels, they long to have a body.'

The aliens' fascination with sexuality, our parental love, our sensuality – all noted by Jacobs and Hopkins – is an expression of their yearning for embodiment. But, at the same time, their disembodiment makes them seem to us closer to angels and, therefore, to God. They are spirits, messengers, intermediaries. We have sensuality, they have spirituality. The hybrid programme, therefore, would be a union of these two virtues – though in what realm this would happen is unclear.

We are, through the aliens, undergoing a spiritual awakening, the effects of which are also, for the moment, unclear. There must be, said Mack, a remembering of the knowledge we have forgotten or lost. As Plato said, we must remember what we knew at birth before we can move on to new knowledge.

'In the end, the abduction phenomenon seems to me to be a part of the shift in consciousness that is collapsing duality and enabling us to see that we are connected beyond the Earth at a cosmic level.'

Mack, as I said, embodies two important alien themes: the cover-up and the third realm. But, in the full context of his thought, it becomes clear that these two themes can also be seen as one. The cover-up is caused by the persistence of Western secular thought in clinging on to its scientist view of the world. We are so attached to this view that we will actively suppress – cover-up – any other. The third realm is an absolute affront to this view; if it exists, then it must displace scientific materialism completely. There can be no repeat of the seventeenth-century deal that separated spirit from matter. Furthermore, there is a sense in which the third realm actively colludes in the cover-up. It is a realm of tricksters, shape-shifters, illusionists, all of whom play with our sense of what is real. Third realm aliens, like their nuts and bolts neighbours, don't want us to know exactly what's going on. In this, the entire capering cast has succeeded triumphantly.

6

Alien Cover-Up

PRESIDENT THOMAS WHITMORE: Sir, regardless of what you may have read in the tabloids, there have never been any spacecraft recovered by our government. Take my word for it. There's no Area 51. There's no recovered spaceship.
ALBERT NIMZICKI: Uh, excuse me, Mr President. That's not entirely accurate.
DAVID LEVINSON: What? Which part?

In *Independence Day* (Roland Emmerich, 1996) Nimzicki, the defence secretary, knows something even the president doesn't. He knows there is such a place as Area 51, that an alien ship was recovered by the government and that alien corpses are being stored there. That the defence secretary knows and the president doesn't is significant. It means government is divided against itself. Some secrets are too important to be entrusted even to the president. The geek, David Levinson, sees the point at once.

Aliens sprang into the contemporary imagination at the same time as a series of atomic explosions announced the start of the Cold War. Flying saucers and mushroom clouds filled the skies at the same moment, a moment of fear, uncertainty and new worlds.

'It is the story,' writes science historian Michael Swords, 'of a secretive phenomenon (UFOs) interfacing with a secretive human activity (military intelligence) at a time of maximum concern and confusion.'

As a result, from the beginning walking into the world of saucers and aliens was like walking into a hall of distorting mirrors. Which, if any, was the real thing?

The Roswell flying saucer crash, to which the exchange from *Independence Day* refers, was the primal incident not just of alien lore, but also of postwar paranoia. The whole sequence of events seemed deliberately designed to induce suspicion. First, the US Air Force confidently announced a saucer had been recovered. Then the authorities equally confidently announced it was a weather balloon. This was naturally assumed to be a cover-up since the alternative – that the USAF couldn't tell the difference between a flying saucer and a weather balloon – appeared to be at least as bizarre as an alien crash. In fact, it really was a cover-up. Later it transpired that it was not a weather balloon at all but an experimental constant-altitude device – a string of balloons – that was designed to listen out for the distant sound of a Soviet nuclear test. Of course, this is now also assumed to be a cover-up.

The CIA was also involved in a serious and very real cover-up, an operation that was to earn it the demonic status it now enjoys in ufological and contactee circles. In 1952 the agency was concerned about the level of public interest in UFOs and the formation of civilian saucer groups. But, deviously, it covered up the level of its concern because it didn't want to further stimulate interest. The reason for this appears to be a fear that a Soviet attack could be launched in such a way that it would be concealed by a UFO flap. 'At the moment of attack,' said one briefing paper, 'how will we, on an instant basis, distinguish hardware (enemy bombers) from phantom (UFO)?' The CIA didn't want to disseminate this possibility. It might give the Soviets ideas.

Also because, again as in *Independence Day*, different branches of government knew different things, it was perfectly possible for officials in senior positions to take the view that the aliens had indeed landed or crashed. J. Edgar Hoover at the FBI certainly thought the air force had a saucer. He was also worried about what had really happened at Roswell. It was an FBI investigation that had discredited the weather balloon story. Government was busily exposing its own cover-ups.

Senator Barry Goldwater seems to have been convinced that something funny was going on. He called the head of Strategic Air Command. In 1994 he told Larry King: 'I called Curtis LeMay and I said, "General, I know we have a room at Wright-Patterson where we put all this secret stuff." I've never heard him get mad but he got madder than hell at me and said, "Don't ever ask me that question again!"'

In 1994 Congressman Steven Schiff also attempted to find out what had happened at Roswell. He was not exactly a believer but he started to become one as he came up against official obfuscation.

'Generally, I'm a sceptic on UFOs and alien beings, but there are indications from the runaround I got that, whatever it was, it wasn't a balloon. Apparently, it's another government cover-up.'

In Canada a secret memo sent by a Department of Transport scientist named Wilbert Smith said it had been learned via the embassy in Washington that flying saucers existed and were 'the most highly classified subject in the United States government, rating higher even than the H-bomb . . . The entire matter is considered by the United States authorities of tremendous significance.'

In Britain scientists and generals secretly drew up a report on UFOs as a result of a wave of sightings around 1950. The six-page document, written in 1951, was used to brief the prime minister, Winston Churchill. The report concluded that all the sightings were explicable by natural phenomena or human activity. Its existence was, nevertheless, routinely denied by the Ministry of Defence for over fifty years. The briefing did not satisfy Churchill. During a further wave of sightings in the US in 1952, he sent some questions to his advisers: 'What does all this stuff about flying saucers amount to? What can it mean? What is the truth?' They remain unanswered.

The conviction of the believers that there is a cover-up in progress is thus reinforced by the suspicion among many in authority that they may well be right. This made for a comical but philosophically resonant state of affairs. Governments could not issue convincing denials because parts of those governments believed other parts were hiding something. That is the point about cover-ups; if they work you know nothing about them. Furthermore, any such denials were confirming instances of the cover-up itself because, of course, if there is no denial there can be no cover-up. In fact, once you believe there is a cover-up in progress, everything becomes a confirming instance.

For the authorities, the trick is, of course, to make it look as though the denial does not emanate from government. This, it is said, was the role of Donald Menzel, who died in 1976. Menzel was the first full-time UFO-debunker. He was an American astrophysicist who wrote or co-wrote three books attacking claims about UFO sightings. Oddly, his own interest in the subject had started with a sighting, which he seems later to have shrugged off. He was enormously influential in discouraging scientific interest in ufology. But, it turned out, Menzel had 'top secret ultra clearance'. He was scientifically involved in 'black ops' (secret projects), though it is not known whether any of these involved UFOs. Also, Menzel's initial involvement with the subject appears to have been on orders from above. His superior at the Harvard Observatory, Harlow Shapley, seems to have specifically assigned him the task of debunking UFO stories. Menzel, if you are a believer, was the non-governmental but actually governmental denier.

But, in any case, even if the denials were sincerely intended, they would be yet further chilling evidence of the depth of the cover-up because even the official deniers hadn't been told the truth. Many became convinced that the aliens had allied with certain humans to form a secret government-within-government that would take over when the time came, hence the genuine ignorance within what we naïvely thought was the actual government. This was occasionally identified as a committee called Majestic 12. Secret papers about MJ-12 were sent to a television producer in 1982. The documents are dated 1952, but they refer back to the establishment of the committee in 1947, the year of wonders. Their authenticity is still bitterly disputed.

'Operation Majestic-12,' they explain, 'is a TOP SECRET

Research and Development/Intelligence operation responsible directly and only to the President of the United States. Operations of the project are carried out under control of the Majestic-12 (Majic-12) Group which was established by special classified executive order on 24 September 1947, upon recommendation by Dr Vannevar Bush and Secretary (of Defense) James Forrestal.'

Roswell is specifically mentioned.

'On 07 July, 1947, a secret operation was begun to assure recovery of the wreckage . . . During the course of this operation, aerial reconnaissance discovered four small humanoid-like beings had apparently ejected from the crash at some point before it exploded. These had fallen to earth about two miles east of the wreckage site. All four were dead and badly decomposed due to action by predators and exposure to the elements during the approximately one week time period which had elapsed before their recovery.'

The saucer was said to be a short-range reconnaissance craft and autopsies on the corpses revealed the aliens were quite different from human beings. The term used to describe them was EBEs – Extraterrestrial Biological Entities.

The alien–government cover-up theory also gave birth to the legend of the Men in Black. These were sinister, black-suited men in black cars who began to appear at the homes of UFO witnesses in the late fifties. They often appeared even before the witnesses had told anyone of their experience. The MIB always warned them against reporting the sighting. Significantly, there is some confusion as to whether the MIB are human or alien or a combination of the two. They were often said to walk strangely and speak mechanically. But the legend, as realized in comics and film, simply shows them to be government agents involved in the concealment from the public of a large-scale alien presence on Earth. Their sudden appearance is reminiscent of the black helicopters associated with cattle-mutilation incidents. These are also subject to conflicting interpretations – are they government or alien or both?

Throughout all this, the problem for the official side was that, with the Cold War hotting up, there were plenty of things to cover up. If, in 1947, they did have a constant-altitude balloon train capable of detecting atomic explosions, then they didn't want the Soviets to know. And, later, when they were testing the Stealth fighter and bomber, the whole point was to keep the technology secret. Both are very strange looking craft whose existence had to be persistently denied. There is speculation that what Kenneth Arnold had, in fact, seen were experimental military craft based on German 'flying wing' designs. It was these designs that eventually led to the configuration of the Stealth bomber.

This type of theory will not go away. The US has since been said to be testing an 8000-mph plane called Aurora which, it is claimed, has been tracked by air traffic controllers at Los Angeles. Aurora is thought by some to be the craft behind many of the large triangular UFO sightings.

There is the further possibility that, as government could not account for the number of UFO sightings, reports of them were dismissed for fear they were actually enemy aircraft. The cover-up was to prevent the public discovering that the Soviets were technologically superior. And so on. Epistemologically speaking, UFOs and aliens were not just a hall of mirrors, they were a minefield.

The additional twist to all this was the later conviction among believers that the aliens could impose amnesia. This made a massive cover-up – by them, not us, or perhaps by them in alliance with some of us – seem feasible. This was, of course, further reinforced by the belief that parts of the government were in collaboration with the aliens. The experiences of Schiff, Goldwater, Churchill, Hoover and Wilbert Smith would seem to add fuel to this particular fire.

This epistemological chaos has left many convinced that more, much more, is going on than we can begin to imagine.

'I am now more than ever convinced,' writes Georgina Bruni in her book *You Can't Tell the People: The Cover-Up of Britain's Roswell*,

> that they are time travellers from our future or another dimension. This would account for why there is a reluctance from our governments to reveal the truth about these encounters. How would you tell the people that there

is an intelligence far more advanced than we are, who are capable of creating such incredible technology? Not only would it be difficult to explain, but also it would open a can of worms that would seriously question history and religion as we know it. We must also consider that since the early 1950s America, Russia and Europe have been struggling endlessly to develop an exotic propulsion system, which they hope to use for future space travel. Are they also experimenting with time travel?

Could be. Who knows? The problem at the centre of the idea of the cover-up is the lack of proof. It is this that creates the fertile soil in which speculation and mistrust flourish. Though they cruise over our cities, abduct our people, deliver momentous messages and warnings, aliens have left no physical proof of their existence, at least none that has satisfied the sceptics. 'Plenty of evidence,' as Nick Pope, a British Ministry of Defence Official says, 'but no proof.' Pope's experiences have persuaded him that there is no government cover-up, but that the aliens are here. This is, in the world of ufology, a distinctly unusual position.

Proof would be an alien body, an alien substance, alien technology beyond anything we now have, an alien craft, anything which we could pass from hand to hand or display in a museum and agree that it constituted incontrovertible proof of extra-terrestrial intelligence. But there is none, at least none available to the public.

(To digress: proof of another realm is a curious matter in its own right. Physical mediumship, which thrived in the nineteenth century, specialized in 'apports' – objects passed from the spirit world to ours. But, since these were all objects, like watches, coins or, occasionally, live animals, that could equally well appear in our world, their mere existence was not enough to convince sceptics. What was required was something that could not occur in our world. One pleasing suggestion I have heard involved two interlocking rings of different types of wood, both without any join. But the spirits have not, so far, obliged. In the case of alien artefacts, the most commonly reported item seems to be a very thin, light metal that does not burn, scratch, break or deform. There are, of course, also corpses. The most ingenious and haunting artefacts are described in Arkady and Boris Strugatsky's novel *Roadside Picnic*. Among the detritus left behind on Earth by untidy alien visitors are: 'empties' – two saucer-sized copper discs fixed about eighteen inches apart with nothing between them but which cannot be squeezed together or prised apart; 'itchers' – small devices which, when activated, cause people in the vicinity to fall into deep depression, freak out or panic; and 'so-sos' – tiny batteries that reproduce themselves. Earth scientists cannot establish how any of these work, nor can they be sure that what they do is what they were designed to do by their alien makers. This last is a very refined and provocative twist.)

But, instead of any such physical proof, all we have is 'evidence', the thousands of accounts of sightings and encounters. To those who deliver the accounts, the aliens have been here – are here – and no more needs to be said.

As Jacques Vallée puts it: 'Through UFO activity, although no physical evidence has yet been found, some of us believe that contours of an amazingly complex intelligent life beyond the earth can already be discerned. The wakening spirit of man, and the horrified reaction of his too-scrupulous theories: what do they matter? Our minds now wander on planets our fathers ignored. Our sense, our dreams have reached across the night at last, and touched other universes. The sky will never be the same again.'

Vallée makes it clear that the lack of physical proof means that the study of UFOs is necessarily indirect. The subject of study is not UFOs themselves, but reports of UFOs. This does not discourage him, though, of course, it does encourage those sceptics who wish to locate all aliens inside the minds of the deranged, the simple and the confused.

In this context, the sceptic is further, if paradoxically, encouraged by the sheer number of reports. If, he reasons, there really have been so many incidents, why isn't there any proof? Surely, during this mass visitation, a few bits of alien litter – as in *Roadside Picnic* – would have been left behind, some shred of an alien spacesuit, the barrel of an alien syringe, a few mangled atoms from an alien disruptor beam. Surely it is inconceivable that the aliens are so conscientiously tidy that, in spite of the millions of humans that yearn and search for proof, they have managed to remove every fragment that might attest to their presence. And, even if they were so tidy, why should they be? Why shouldn't they let us know? Or, why should they care whether we know or not?

For the believer, there are, broadly, two possible answers. Either the aliens don't want us to know or our governments don't want us to know. In the latter case, there is no absence of proof, the wrecked ships and corpses are simply kept locked away from the eyes of the masses. If it's the aliens who don't want us to know, that may be because of all the reasons involving their interbreeding programme. Certainly, according to some theories, they are very ingenious in their cover-ups. Not content with merely inducing amnesia, they exploit human culture. In 1993, for example, Nick Pope was analysing a wave of sightings that had happened on 30 and 31 March. The dates were the same as some earlier sightings.

'Over and over again,' he writes in his book *Open Skies, Closed Minds* (1998), 'I pondered the significance of the date. The odds against such a phenomenon occurring coincidentally on the same night three years apart are high. That suggests that the date was not random, but was deliberately chosen and planned. Furthermore, it was chosen by an intelligence familiar with human frailties. Newspaper reports of incidents occurring that night would run on 1 April, the day when every national and many provincial papers carry an April Fool story.'

People would simply laugh – 'Isn't this exactly the reaction an alien force might hope to achieve by capitalizing on a time when the world is unreceptive, when everyone expects bizarre stories and dismisses them out of hand?'

But, whatever games aliens or governments might play, there is another motive for cover-ups, which, though it could not realistically be claimed to account for the total effectiveness of the concealment, is, nevertheless, a very effective instrument of censorship. This is embarrassment.

A US survey of UFO press coverage conducted in 1970 by

journalism professor Herbert Strentz found that fear of ridicule was the second most common reason for not reporting an incident. Strentz also found that 18.4 per cent of newspaper reports contained evidence that witnesses had been ridiculed or harassed for making their reports public. There is what is commonly known as 'the giggle factor'. Mention your close encounter in polite society and people will think you have gone mad. I know – people have backed away from me when I told them the subject of this book.

There is also a class element to this. An episode of the US sitcom *Dharma and Greg* shows Greg ridiculing the entire topic of aliens as a subject of interest only to people who drive pick-up trucks down dirt roads – that is to say, rural hicks. This may well have been a direct reference to the Travis Walton abduction case. The point is that extraterrestrial concerns of any kind are a massive social disadvantage. SF or UFO conventions are generally regarded as colloquies of sad, male nerds. An attender at one such conference was overheard to remark; 'It's going really well, a girl came last month.'

The character of Brandon in the film *Galaxy Quest* (Dean Parisot, 1999) is a nerd. The cast of a *Star Trek*-like TV show discover a real alien race has taken them seriously and built a working replica of their non-working fictional ship *Protector*. Brandon knows every nut and bolt of *Protector* and, secretly in his bedroom, believes the show is real. At a decisive moment, his intimate knowledge of the fictional *Protector* saves the crew of the real *Protector*. His nerdishness is touchingly vindicated. The fact that both Brandon and the dishevelled world of the fan convention are so readily evoked by the movie indicates how fixed in the popular mind is the link between aliens and social incompetence.

Such prejudice is further reinforced by the way stories of alien encounters predominantly appear in the least credible and most downmarket contexts. Indeed, in America supermarket tabloids seem to have the genre more or less to themselves. One gloriously paranoid theory about this is that the government is actually using the tabloids to discredit UFO and alien stories. The idea is that, if it appears there, it must be false. This theory has some circumstantial backing in that Generoso Pope, the proprietor of the *National Enquirer* in its glory days as the leading US tabloid, had, in fact, been a CIA intelligence officer and is thought by some to have continued to do the government's bidding as a press baron. We shall probably never know, Pope died in 1988.

The movie *Men in Black* (Barry Sonnenfeld, 1997) notes the significance of tabloids to alien lore when Agent K picks up these papers, explaining they are the most reliable sources of news about alien activity. The joke is that what the tabloids report is, in fact, true. This doesn't matter either because nobody believes them or because only unimportant people believe them. The truth is in plain sight, but it is a vital aspect of the work of the Men in Black to conceal this truth in order to prevent public anxiety.

'There's always an Arquillian battle cruiser,' says Agent K, 'or a Korilian death ray, or an intergalactic plague that is about to wipe out life on this miserable planet. The only way these people get on with their happy lives is they do not KNOW ABOUT IT.'

John Mack's story also involves embarrassment. Plainly the Harvard authorities were embarrassed to find a senior member of their staff was a believer in alien abductions. Academic embarrassment is, indeed, a powerful agent of suppression. Many believers compare the refusal of academia to take UFOs seriously to the reaction of the French Academy of Science in the eighteenth century to reports that stones had fallen from the sky. Such things, it was said, were impossible. The French, like modern Americans, made the scientifically fatal assumption, 'It can't be, therefore it isn't.' The stones were, of course, meteorites.

Jacques Vallée quotes a restricted newsletter from the Smithsonian Astrophysical Observatory on the subject of UFOs. The paranoid tone is a remarkable expression of the 'It can't be, therefore it isn't' ideology.

'It is exceedingly undesirable to become associated with these "sightings" or the person originating them . . . On no account should any indication be given to others that a discussion even remotely concerned with UFOs is taking place.'

The subject is socially and academically toxic. This would appear to confirm John Mack's view that it overturns a belief system that is too precious for its adherents to let go. Indeed, according to philosopher Michael E. Zimmerman, this seems to be why it is so embarrassing to academics. Abductions, he says, amount to 'forbidden knowledge of hidden events' and these events 'conflict so sharply with accepted views about "reality" that the event can scarcely be brought up in polite society, much less made an object for publicly funded research'.

Aliens are also professionally toxic. Airline pilots started out by regularly reporting UFO sightings, but, after a time, it became clear that this was a risky thing to do. Bureaucratic hassle and behind-the-back mutterings about their sanity persuaded many to cease filing reports.

'If I saw a flying saucer flying wing-tip formation with me,' said one anonymous pilot interviewed by Captain Edward Ruppelt, head of the Wright-Patterson UFO project in 1953, 'and could see little men waving – even if my whole load of passengers saw it – I wouldn't report it to the air force.'

In fact, the embarrassment factor flows smoothly into the government cover-up theory. After the first official reactions to Arnold and Roswell and the report of Project Sign, which gave credence to the extraterrestrial hypothesis, the USAF and government ruthlessly set out to put an end to the entire UFO flap. Air Force chief-of-staff General Hoyt S. Vandenberg famously rejected Sign's 'Estimate of the Situation' out of hand. It was ordered declassified and destroyed. For a while the USAF denied it had ever existed. It was just too embarrassing.

The problem with the programme of denial that ensued was that it was so staggeringly inept that it confirmed even sceptical suspicions that a cover-up was indeed in progress. In January 1953, for example, the USAF and the CIA convened the Robertson Panel. This was a group of four academics. They spent just twelve hours examining all the data of the previous six years of UFO sightings and blithely concluded that most had conventional explanations and the rest were not worth the bother of further investigation. This stunningly off-hand piece of work transformed government policy and downgraded all further official interest in UFOs. Projects Grudge and Blue Book, the successors to Project Sign, became explicitly debunking operations.

The Robertson Panel, however, could be excused on the basis

that its purpose had been 'to evaluate any possible threat to the national security posed by Unidentified Flying Objects'. Arguably, even if the saucers were real, they might not actually threaten the United States. The Robertson conclusions, however, went a little further. These said there was no threat to the US and added, 'there is no evidence that the phenomena indicate a need for the revision of current scientific concepts'. As a result, they recommended that 'national security agencies take immediate steps to strip the Unidentified Flying Objects of the special status they have been given and the aura of mystery they have unfortunately acquired'.

The final paragraph is a small masterpiece – the word 'inimical' is especially fine: 'We suggest that this aim may be achieved by an integrated program designed to reassure the public of the total lack of evidence of inimical forces behind the phenomena.'

But mere national security would provide no excuse for perhaps the most hilariously compromised official attempt at denial or, if you prefer, cover-up. This was the Scientific Study of Unidentified Flying Objects, commissioned by the USAF and directed by Edward U. Condon, professor of physics and astrophysics at the University of Colorado. Ironically, the Condon report was welcomed by many because it transferred responsibility for UFO investigation from the military to civilian scientists, the implication being that it would be a much fairer assessment than any that had been done by the irately sceptical military. Even J. Allen Hynek, later to become the leading UFO believer of his day, welcomed the formation of the Condon committee.

'What a pleasure it was,' he said after a dinner meeting with members of the committee, 'to sit down with men who were open-minded about UFOs, who did not look at me as though I were a Martian myself.'

The pleasure was to be fleeting. The committee sat between 1967 and 1968. It suavely concluded that 'further extensive study of UFOs probably cannot be justified in the expectation that science will be advanced thereby . . . UFO phenomena do not offer a fruitful field in which to look for major scientific discoveries'. For a number of reasons, this persuaded absolutely nobody and probably did more to convince people a cover-up was going on than any other single document.

For a start, one third of the cases covered by Condon remained unexplained and, it was clear, the committee wasn't interested in trying to explain them. One tricky sighting was said to be 'almost certainly a natural phenomenon, which is so rare that it apparently has never been reported before or since'.

Furthermore, almost from its inception, the Condon report had self-destructed. Even before the committee began its deliberations, Condon himself had made his feelings clear: 'It is my inclination to recommend that the government get out of the UFO business. My attitude right now is that there's nothing to it, but I'm not supposed to reach that conclusion for another year.'

As if that wasn't enough, Condon's project director Robert Low wrote a confidential memo to his boss, again before the committee even sat.

'Our study would be conducted almost exclusively by non-believers who, although they couldn't possibly prove a negative result, could and probably would add an impressive body of evidence that there is no reality to the observations. The trick would be, I think, to describe the project so that, to the public, it would appear a totally objective study, but, to the scientific community, would present the image of a group of non-believers trying their best to be objective, but having an almost zero expectation of finding a saucer.'

No doubt about it, this was a cover-up.

Condon and the ensuing revelations about its predetermined bias convinced believers, the open-minded and even the mildly sceptical that there was no chance of any serious consideration of the UFO problem by the authorities. A cover-up was in progress either because government really didn't want us to know the truth or because they had simply made up their minds in advance that the entire UFO phenomenon was a case of mass hysteria. One effect was to privatize all future UFO research. Around the world private investigative groups sprang up to do what the authorities had so signally failed to do – find out what exactly was going on.

A further effect was to disgust many scientists. James E. McDonald, professor of atmospheric sciences at Arizona, spoke at a symposium on UFOs sponsored by the American Association for the Advancement of Science in Boston in 1969, not long after Condon.

'No scientifically adequate investigation of the UFO problem has been carried out during the entire twenty-two-year period between the first extensive wave of sightings of unidentified aerial objects in the summer of 1947 and the convening of this symposium . . . I believe science is in default for having failed to mount any truly adequate studies of this problem.'

He added that the Condon report did not offer 'anything superior to the generally causal and often incompetent level of analysis' that had been the distinguishing feature of previous USAF reports. McDonald passionately believed that something strange was happening and seized every opportunity to publicize his conviction. He became the believer's saint, constantly at war with their demon, sceptical investigator Philip J. Klass. He lost. He was removed from a navy research contract because of his beliefs and, in March 1971, he was embarrassed in front of a congressional committee. He was testifying about the atmospheric impact of supersonic passenger planes and his UFO interests were used to discredit his testimony. Aged fifty-one, he killed himself on 13 June of that same year.

Hardened conspiracy theorists, of course, don't believe he did kill himself. Leo Sprinkle, psychologist and UFO researcher, has claimed that people had been killed to protect the government's alien secrets and McDonald was felt by many to be one of these. Even the very rapid death of Congressman Steven Schiff from cancer, soon after he had started asking questions about Roswell, is seen as suspicious.

But perhaps the most important figure to be disgusted by official attitudes to UFOs was J. Allen Hynek. Hynek, who died in 1986, is authority's worst nightmare, the expert who goes native. He was appointed by the USAF in 1948 to look into UFOs. He saw there were problems with explaining many of the sightings, but, on the whole, he quietly did his work, seeking out conventional explanations. In 1966, however, he investigated a series of cases in Michigan. One explanation for some of them, he told the press, was marsh gas (methane). The idea that UFOs were, in fact, what is more generally known as swamp gas was just too absurd, an excessively transparent groping for any old explanation. It became a national joke. In *The X-Files* sceptical Scully defends swamp gas as a legitimate explanation in the face of Mulder's scorn. Hynek could keep quiet no longer.

In fact, he had never been quite convinced by the government line. His work on the Grudge report, issued at the end of 1949, involved analysing 237 of the best sightings. Around 32 per cent could be explained astronomically, 12 per cent were balloons and 33 per cent were aircraft, hoaxes or too sketchy to evaluate. That left 23 per cent classified as 'unknown'. As Hynek was later to point out, this was a staggering figure, but the press did not pick it up. Seemingly the USAF's debunking of the story had, by that time, become very effective.

But it was not until the sixties that Hynek's view of UFOs really began to change under the influence of, among others, Jacques Vallée. In 1972 Hynek published *The UFO Experience: A Scientific Approach* and, in 1973, he founded the Center for UFO Studies (CUFOS) at Northwestern University. In 1977 came *The Hynek UFO Report*. His reputation as the government officer who defected to tell the truth was established.

Hynek is variously described as crazy, arrogant, brilliant, wise, humble etc. He is, in short, a man about whom people felt it necessary to have opinions. But, whatever else he was, he was certainly systematic.

It was his careful categorization of alien encounters which was to make his work famous far beyond the devotees and debunkers of alien lore. For the weakest contacts, he devised three self-explanatory categories: nocturnal lights, daylight discs and radar sightings. For stronger, closer contacts, he added three more categories. A Close Encounter of the First Kind, he said, is where a UFO is simply seen. A CE of the Second Kind involves a degree of interaction between the UFO and the environment and the witness. A CE of the Third Kind involves an encounter with the aliens themselves.

Close Encounters of the Third Kind was, of course, the title of the Steven Spielberg film that came out in 1977, the same year as *The Hynek UFO Report*. There seems to have been some friction between Hynek and Spielberg over the use of his material. It may well have been resolved by Hynek's cameo performance in the film. At the landing site, when the alien mother ship descends, he can be seen smoking a pipe amidst the scientists and technicians. As the character of Claude Lacombe, played by François Truffaut, is modelled on Jacques Vallée, the film amounts to a fictional snapshot, complete with real characters, of the factual UFO wisdom of the time. And it is, of course, a cover-up film. Early on an airline pilot is shown declining to report a sighting out of embarrassment. And, when the government does make contact, its first instinct is to keep the people away. They are foiled in this by the fact that the aliens have specifically invited ordinary people to the landing site by, apparently telepathically, giving them a vision of the place.

Hynek's importance is not really in the development of UFO and alien lore. He remained cautious to the end, concentrating primarily on analysing and categorizing sightings, though he did start to give some credence to contact and abduction stories. But his defection from the government line, his establishment of CUFOS and his role in inspiring perhaps the most persuasive alien contact film of all make him the pivotal figure in the conflict between sceptical government agencies and believers. CUFOS has more than 50,000 cases of UFO sightings and experiences on its files. Hynek was once asked to provide evidence for the existence of UFOs. He answered, with an implicit gesture at this mountain of documents, 'Where do you want the truck to stop?'

The problem is that, big as the mountain may be, this remains evidence and not proof. Believers often point out that murderers are often convicted on less evidence than is contained in a couple of well-witnessed UFO sightings. But, for sceptics, murders are relatively likely, flying saucers are not.

Furthermore, almost all the evidence can be interpreted in wildly different ways. Just how far this can go is shown by a crucial story published in the *New York Times* on 14 January 1979. This was about CIA papers that had been released as a result of a Freedom of Information Act lawsuit. The *NYT* story was buried on page 23, but it seemed sensational. The papers showed that 'the agency is secretly involved in the surveillance of unidentified flying objects and has been since 1949'. The documents certainly exposed a lie – the CIA had previously said it had 'closed its books' on saucers in 1952 – but what did they actually prove? The *NYT* also quoted William Spaulding, head of the civilian group Ground Saucer Watch: 'After reviewing the documents, Ground Saucer Watch believes that UFOs exist, they are real, the US Government has been totally untruthful and the cover-up is massive.'

Terry Hansen in his book *The Missing Times: News Media Complicity in the UFO Cover-Up* also treats this story as a vital revelation of the true depths of the government cover-up. In this, of course, they were supported by the fact that the CIA had gone to some lengths to keep the documents secret. This interpretation seemed to be confirmed by a *Washington Post* article five days later with the startling intro: 'During two weeks in 1975, a string of the nation's supersensitive nuclear missile launching sites and bomber bases were visited by unidentified low-flying and elusive objects, according to Defense Department reports.'

Debunker Philip J. Klass, however, starts from the same newspaper reports and concludes that what the CIA papers actually reveal is their almost total lack of interest in UFOs. He analyses the 879 pages released and points out that fewer than 350 of those that had formerly been classified cast any light on the CIA's interest in the saucers. This was over a thirty-year period so 'CIA officials and employees had written an average of only one page of classified UFO-related material per month'. Many of these pages were duplicated memos and many contained secondhand reports of sightings from the US and the Soviet Union. Overall, Klass concludes, the CIA didn't much care about UFOs, but, insofar as it did, it was concerned only that they be used as a cover for a Soviet attack.

As Klass and Hansen make clear, the same thing looks different when viewed from different angles. This was even more true of physical evidence. NASA, for example, interpreted glowing objects filmed around the space shuttle *Discovery* while in orbit as 'orbiter-generated debris illuminated by the sun against a dark background'. Probably they were water ice. Jack Kasher, physics professor at the University of Nebraska, looked at the same film and produced five separate mathematical proofs that the water-ice explanation was impossible. He thought the images were physical evidence of UFO activity in low earth orbit. Was NASA covering something up? Certainly, according to *The Right Stuff* (Philip Kaufman, 1983) John Glenn, the first orbiting American, saw glittering sparks outside his craft. They appeared to save him in a moment of crisis. The film does not suggest an alien explanation but poetically implies that the sparks rose from the fires of chanting Australian aborigines. Either way, it's nothing like the official story.

Beyond all this lies the question of motivation: that of the government, the debunkers, the believers and even the aliens themselves. What did each of them gain or lose from a cover-up? Or, to put it another way, why a cover-up?

In part, I have suggested some answers to this question already. The government covered up because it was hiding something – its own weapons, fear of Soviet technology, a deal with the aliens or its own uncertainty about what other departments knew. There were further sub-categories. George Adamski's Venusian Space Brothers, for example, had the secret of free energy. The government could not let this get out because it would bankrupt the oil companies. It was much the same kind of reasoning that argued the Americans were hoarding alien technology to give them a military advantage. On top of this there was the question of disinformation, the cover-up of the cover-up. Many thought Adamski-type stories, like the supermarket tabloid tales, were planted to discredit alien speculation.

The debunkers, meanwhile, are either crusaders for the truth – Klass quotes Pasteur, 'The greatest derangement of the mind is to believe in something because one wishes it to be so' – or they are tools of the government or, in Mack's terms, they are ferociously defending a dying world view.

The believers are also crusaders and they have the additional incentive to believe in a cover-up because it explains the lack of physical evidence. They also, for Klass, seek to give meaning and significance to their little lives. They are, by and large, passionately mistrustful of governments and assume the existence of multiple conspiracies – not, in the light of the ineptitude of most of the attempts at cover-up, an unreasonable assumption.

The aliens, meanwhile, have a programme of hybridization which they need to keep quiet for fear of disturbing humans or because it is an aspect of a project, the exact nature of which we are not yet ready to understand.

But there is a wider context involved which is really a way of answering the second question: why a cover-up? The answer is that covering up is what we, in the postwar world, do.

At one level, the reason for this is that identified by Michael Swords, 'a secretive phenomenon (UFOs) interfacing with a secretive human activity (military intelligence) at a time of maximum concern and confusion'. But this says a little too much a little too quickly. It requires explanation.

As a result of the technological take-off and the accompanying dizzy sense that anything was possible in the late forties, there was a lot more to hide and many more reasons to suspect that things were being hidden. This, combined with the global bipolarity that emerged after 1945, meant that the fate of the whole world depended on secrets held by just two parties, the Soviets and the US. The key secret was nuclear power, in the form of the atom bomb and, later, the hydrogen bomb. This was a qualitatively different secret from any that had gone before. Previous secrets – radar, Britain's breaking of the Enigma code – had specific tactical significance. But nuclear power had total strategic significance. It did not just alter war, it changed its fundamental nature. Though it was not true until much later, when nuclear stockpiles had reached surreal proportions, it seemed, from the moment the Hiroshima bomb exploded, that these weapons, if used, would destroy the whole world.

Of course, the Soviets and others soon had nuclear weapons, so that secret was out. But a climate had been created in which the idea of decisive secrets was familiar. Governments were expected to be concealing incredible, world-transforming knowledge. Nuclear weapons had been such a massive change that, naturally enough, people assumed that further changes on a similar scale could be on the way. The possibilities that the

Soviets were developing flying saucers or that aliens were co-operating with government in developing new weaponry were not, therefore, easily rejected, even by the most sceptical.

If the US government wanted to keep its secrets to prevent a Soviet attack, then, on the whole, the American people would probably be sympathetic. This, of course, presupposes a degree of trust in government. People would assume it was not keeping unnecessary secrets or ones that would threaten the lives of Americans. When the UFO flap started in 1947, this was, in fact, the prevailing attitude. As polls showed, most people thought the saucers were Cold War technology belonging to one side or the other. The idea that these were extraterrestrial visitations was either a specialized, if quite widespread, theory or a source of entertainment.

But, as the alien theory began to take hold, it was accompanied by the conviction that the government was hiding something that was of more than military significance. An alien visitation should not just be an occasion for improving one's defences against the Soviets, it should be a, if not the, decisive moment in human history. In fact, there seems to be some strange symbiotic relationship between the concepts of extraterrestrial life and the cover-up. The scientist Gil Levin still insists, for example, that the experiment he put on board the Viking Mars lander in the seventies did prove the existence of Martian life. NASA denials of this naturally confirm a cover-up.

The idea of the alien cover-up is a precursor for what was to become a fundamental change in the level of trust between people and government. It anticipates the later mania for cover-ups as exemplified in the US by the conviction that the true assassins of President Kennedy had not been caught, not to mention Iran-Contra and Watergate, and in Britain, by the Weapons of Mass Destruction issue following the Second Gulf War.

As anthropologist Charles A. Ziegler puts it: 'The anti-government sentiment within the UFO subculture found no echo in the larger national culture until the late 1960s. From that time onward, however, a series of events such as the Watergate and Iran-Contra affairs eroded public confidence in government, a trend that placed the public's views about the trustworthiness of government on a converging course with the views of the ufologists.'

But, in fact, the big cases are just high-profile examples of what has become a normal way of conducting our affairs. We now assume that the government routinely covers things up and that it is our right and duty to uncover such things. Individuals, meanwhile, are held to account for their sex lives and financial dealings, however unremarkable they may be. We live in a world of perpetual cover-up and periodic uncovering.

The idea that the government was covering up contact with aliens was a moment at which we caught a virus of mistrust. This bug spreads rapidly. Moreover, it mutates, creating ever more elaborate versions of itself. In normal circumstances this mutation process is stopped by the act of uncovering. But the aliens have never been uncovered. The virus has thus been allowed to mutate unchecked. In fact, its mutation rate has been accelerated by the discovery of what is being covered up. For example, in *Independence Day* President Thomas Whitmore says there is no Area 51. And, officially, he is right. It doesn't exist. It has never appeared on aviation charts or US geological survey maps. It must be a fantasy of the paranoid imagination. Except that it isn't. Area 51 exists.

It is at Groom Dry Lake in Southern Nevada. Nuclear weapons had been tested round here since 1951 and construction of Area 51 itself began in the mid-fifties as a site for testing the U-2 spyplane. It then went on to be used for the development of Black Budget projects (secretly funded so as not to appear in public documents) like the A-12 and SR-71 Blackbirds, the F-117 Stealth fighter and, it is rumoured, the hypersonic Aurora. The call sign of the Area 51 control tower is – this is too good to be true – Dreamland and this is how it is commonly known in alien lore. To flyers it is known as the Box – airspace over Dreamland is fiercely restricted. The base has also been used for many other secret military activities including work on the Strategic Defense Initiative (Star Wars) and commando training. Unmarked Boeing 747s have been seen taking off from Las Vegas, 100 miles to the south, and from Palmdale, California, home of Skunk Works, the headquarters of Lockheed Advanced Development Projects.

These take up to 2000 employees into Dreamland. Their security oath prevents them from even mentioning the place. Punishment for violations of this go up to ten years in prison and a fine of $10,000.

In 1984 the USAF seized control of 89,000 acres of land round the base. A challenge to this move in Congress was met with the statement that it was done on higher authority. Representative John Seiberling nobly said that 'there is no higher level than the laws of the United States'. But the fact that he appeared to be wrong reinforced the sense that Area 51 was outside all democratic control. In addition, of course, the official position remained that it did not exist so no land could have been seized.

This position was somewhat compromised by the fact that the seized land did not include White Sides Mountain, which overlooks the base. From here photos of the non-existent base were taken and widely circulated. The air force moved to grab this land as well. But then an even better site was discovered by self-styled psychospy Glenn Campbell who went on to produce his 'Area 51 Viewer's Guide'. Conflict between the authorities and the curious became a more or less permanent feature of the land around Area 51 with security personnel known as the Cammo Dudes – because of their camouflage gear – constantly rounding up watchers who get too close. In fact, almost anywhere is too close since if you saw the base at all you were in the risky business of exposing a government cover-up.

That much is known. What is not known, obviously, is what exactly goes on at Area 51. In a trusting world, it would be assumed that everything that happened there was for the improvement of the security of the American people. But this is not a trusting world.

The basic theory – and it is the one exemplified in *Independence Day* – is that the alien corpses and technology recovered after the Roswell crash are kept at Area 51. The Roswell incident had, in fact, more or less been submerged by the blizzard of USAF denials after the initial announcement that a disc had been recovered. For the next thirty years it barely featured in alienology. Then, during the seventies, one Leonard H. Stringfield had given a series of lectures about alien bodies and saucer wreckage. Stringfield also suggested that the Frank Scully book *Behind the Flying Saucers* had not, in fact been a hoax. Rather, it had been discredited as part of a (here we go again) government cover-up. But the full resurrection of Roswell began in 1978 when ufologist Stanton Friedman met L. W. Maltais. Maltais had been told by his friend Grady L. Barnett, who died in 1969, that he has seen both the saucer and the alien bodies. Friedman subsequently worked with Charles Berlitz and William L. Moore to produce the book *The Roswell Incident* in 1980. From that point onwards, Roswell was to be the central alien story of our time.

In fact, the story that the Roswell corpses were kept at Area 51 may itself be a cover-up. According to David Darlington in his book *The Dreamland Chronicles*, there is another theory that the recovered saucers actually came from Kingman, Arizona, and they're kept at Papoose, south of Area 51 at Area S-4:

Since 1953 a satellite government outside the control of Washington, D.C., has pursued this matter in a secret program (which may or may not be descended from the fabled 'Majestic 12' committee, allegedly mandated by President Truman) headquartered at Los Alamos, New Mexico, directed by Dr Edward Teller, and created by then Vice President Richard Nixon. Moreover, this program has been conducted not by our species alone, but with the limited cooperation of extraterrestrial biological entities ('EBEs') from the Zeta Reticuli star system – who, incidentally, are not slimy and reptilian, but answer to the classic description of the 'typical' gray alien: hairless and roughly four feet tall, with big, dark, wraparound eyes.

For unexplained reasons these aliens, known to employees as 'The Kids', require the element boron which is found throughout the American southwest. That explains why places like New Mexico and Arizona are such alien hotspots.

Darlington's attitude to this scenario is that it's a 'heck of a yarn' and is somewhat less sensational than other claims about Area 51. These include aliens who are here for our DNA and aliens who are not extraterrestrials at all. In fact, they are beings manufactured at Area 51 'as part of an elaborate hoax designed to enslave the world population before a bogus "extraterrestrial" threat'. This may be the work of an evil illuminati, a group possessed of arcane knowledge.

Evidence of the cover-up is provided from time to time by people claiming to have been inside Areas 51 or S-4. The most notable of these is Bob Lazar, meeting whom, according to Darlington, is, in ufology circles, 'tantamount to meeting Bob Dylan'. In 1989 Lazar appeared on a local TV show in Las Vegas and claimed that, on the recommendation of Dr Edward Teller, he had been hired to work at Area 51. In fact, when he got there he was taken instead to S-4 where he was assigned to look into the propulsion system of alien craft. On entering, he had seen a saucer about thirty-five feet in diameter. Lazar also discovered from documents he was given that sixty-five genetic alterations had been performed on humans in the course of our evolution and that three spiritual leaders, including Jesus Christ, had been made by alien engineers.

Lazar's revelations turned the little town of Rachel, near where Lazar said he'd watched test flights of the saucers, into a place of international pilgrimage. A black mailbox, nineteen miles south of Rachel and in the middle of nowhere, became the gathering place for pilgrims. And the Rachel Bar and Grill renamed itself the Little A-Le-Inn.

Roswell and Area 51 live on in TV shows, advertisements and in the cultural vernacular of the young. They both give sacred locations to the quasi-religious contemporary devotion to alien lore and they are emblematic of the pervasive conviction that the government no longer covers things up for our own benefit, but now does so for its own dark purposes. Or, indeed, it covers things up as a result of a malign contractual relationship with the aliens.

Extraterrestrials were the first – and they remain the most vivid – expression of the postwar cover-up culture. This is a culture of discomfort, fear, uncertainty and loss of the apparent anchor of the self. It is the culture of alien abduction. The aliens emerged from the depths of a widening abyss between government and people, between those thought to be in the know and those fearful that they are not. And since, for the moment, there is no prospect of that abyss ever narrowing to its former proportions, the aliens are here to stay.

7

Marshall Applewhite

On 26 March 1997, thirty-nine members of the Heaven's Gate cult, including their leader, Marshall Herff Applewhite, were found dead in a house in San Diego. The corpses were on bunk beds, clad in black trousers and tops and each draped with a square of purple cloth. Eight of the men, including Applewhite, had been castrated. Thirty-seven had died after taking a mixture of phenobarbitone and vodka. The last two had tidied up before taking opiates and putting plastic bags over their heads. They all died convinced that the time had come to leave their earthly bodies and join their alien saviours on board a spaceship that was trailing the Hale Bopp comet, then clearly visible in the night sky. One of the dead was Thomas Nichols, the 59-year-old brother of Nichelle Nichols, the actress who played Lieutenant Uhura, the communications officer, in the first series of *Star Trek*.

Heaven's Gate dated back to 1972 when Applewhite, the Texan son of a Presbyterian minister, met Bonnie Lou Trousdale Nettles in a Houston mental hospital where she was a nurse and he a patient. He was, it seems, bisexual and wished to be cured of his gay impulses. He was a divorced college music professor and she a married mother of four. She left home to travel with Applewhite. They were arrested in 1974 in Herlingen, Texas. He was charged with stealing cars, she with credit-card fraud. In spite of which, they were jointly acquiring a small, messianic reputation. Erik Davis in his book *TechGnosis* (1998) links their success at this point to the disappointed dreams of the sixties generation.

'Heaven's Gate began in the early seventies, when a wave of flying saucer cults zoomed into the frazzled spiritual vacuum that followed the collapse of countercultural utopia.'

Where earthbound dreams had failed, maybe extraterrestrial ones would succeed. Applewhite and Nettles formed a group known successively as Next Level Crew, Total Overcomers, Human Individual Metamorphosis and, eventually, Heaven's Gate. Throughout, the core of their faith seems to have been the conviction that an extraterrestrial being would arrive to deliver them to a higher plane of existence. Many cults are relaxed, harmless social groups with ever-open doors and few demands placed upon their followers. But Heaven's Gate was a 'high demand' operation. Applewhite, by then known as 'Do', explained what was required of its adherents on the Heaven's Gate web site.

> Leaving behind this world included: family, sensuality, selfish desires, your human mind, and even your human body if it be required of you – all mammalian ways, thinking, and behavior. Since He had been through this metamorphic transition Himself from human to Level Above Human – under the guidance of His Father – He was qualified to take others through that same discipline and transition. Remember, the One who incarnated in Jesus was sent for one purpose only, to say, 'If you want to go to Heaven, I can take you through that gate – it requires everything of you.' Do is doing Jesus' work.

It seems that an 'away team' from an 'evolutionary level above human' had, from the 1920s to the 1950s, 'picked and prepped' the human bodies required for their task. They actually came into those bodies in the 1970s. They entered the admiral and captain – Applewhite and Nettles – first. From 1975 onwards they rounded up the rest of their crew. A message from a 'presently incarnate' extraterrestrial explains the process. This also accounts for the Roswell crash and similar incidents.

> In the early 1970's, two individuals (my task partner and myself) from the Evolutionary Level Above Human (the Kingdom of Heaven) incarnated into (moved into and took over) two human bodies that were in their forties. I moved into a male body, and my partner, who is an Older Member in the Level Above Human, took a female body. (We called these bodies 'vehicles,' for they simply served as physical vehicular tools for us to wear while on a task among humans. They had been tagged and set aside for our use since their birth.)
>
> It seems that we arrived in Earth's atmosphere between Earth's 1940's and early 1990's. We suspect that many of us arrived in staged spacecraft (UFO) crashes and many of our discarded bodies (genderless, not belonging to the human species), were retrieved by human authorities (government and military).
>
> Humans with deposits containing souls can likely be identified at this time as some of those who are rapidly losing respect for this world or its 'system.' They are, from the establishment's point of view, being irresponsible or anti-social – and will be seen by the world as duped, crazy, a cult member, a drifter, a loner, a drop-out, a separatist, etc.

In 1975 Applewhite and Nettles had arrived in the San Francisco Bay Area and given presentations to audiences of several hundred at Stanford University and various other colleges. Now the pair were known as 'Bo' and 'Peep'. In October a plan to appear at the San Jose Public Library before an audience of 500 was abandoned because of increasingly bad publicity. There had been reports from Oregon, where the couple had previously been, of twenty or more people vanishing as a result of the 'UFO cult'. They were forced to leave the Bay Area, but, by now, they led a crew of 200.

There were, at this point, clear similarities to The People's Temple, the Christian group led by James Warren Jones. This, too, had attracted a hard core of several hundred followers and, in the process, had a few run-ins with the authorities, notably when the government began to investigate Jones's claimed cures for cancer, heart disease and arthritis. The Temple was also a peripatetic movement and also periodically subject to press exposure. Its belief system, like that of Heaven's Gate, became progressively more florid and extreme. Jones came to believe in 'Translation' in which he and his followers would all die together and be taken to another planet for a life of perpetual bliss. Eventually Jones and more than 900 of the Temple's membership moved to Guyana, where they established Jonestown. There, on 18 November 1978, after Congressman Leo Ryan, who had been investigating the cult, and four others had been killed by Temple security guards, all 913 of them died, most from a cocktail of grape juice and cyanide combined with various sedatives.

Conspiracy theories, similar to those involved in alien cover-up theory, circulated around the Jonestown mass suicide. Usually, these assumed the crime was not suicide but mass murder. The CIA was said to be using the Temple to experiment with methods of mind control. This, it was said, had been uncovered by Congressman Ryan and everybody had to be killed to preserve the secret. Another government assassination theory suggested the people were killed because they threatened capitalism with their display of communal living. Jones had, indeed, been veering towards communism and some believed Jones was an ally of the Soviet Union and Cuba.

Jonestown led to a widespread anti-cult movement. The idea was to spot the dangerous self-destructive symptoms in advance. Heaven's Gate, however, was not spotted. It had continued its peripatetic existence. In 1989 Nettles had died, leaving Applewhite in sole charge. But, by 1997, the numbers of his followers were dwindling and, as an autopsy later revealed, Applewhite was suffering from serious heart disease. Many say these two factors, rather than any clear belief in the alien space-ship, were what motivated his decision to die at that moment.

Plainly, there are broad historical points to be made about

both Heaven's Gate and The People's Temple – the apocalyptic mood of the time of their inception, the Vietnam War, the residual hippie conviction that the masses were systematically being denied a better, more fulfilled life by government, the American tradition of peripatetic preachers and highly diverse religious commitments and so on. For my purposes, however, the most significant point is that both cults were driven by a conviction that salvation lay beyond the Earth. In both cases devotees were to be taken to extraterrestrial locations.

This places them both in the postwar extraterrestrial tradition. Heaven's Gate was the more firmly entrenched in ufology. Its followers were said to be *Star Trek* fans and the outfits they were dressed in at their deaths looked remarkably like the ceremonial garb of the Vulcans, the benign aliens in the TV series.

In addition, the idea of a space version of Noah's Ark, sent to save the elect of Earth, can be found within the mainstream tradition of SF. In Arthur C. Clarke's *Rendezvous with Rama*, first published in 1973, soon after Applewhite and Nettles had got together, a huge alien ship, *Rama*, arrives in the solar system. It is empty but it can support human life. The 'Cosmo Christer' – Christian – on board the ship sent to investigate *Rama* suddenly realizes its significance.

'Our faith has told us to expect such a visitation, though we do not know exactly what form it will take. The Bible gives hints. If this is not the Second Coming, it may be the Second Judgment; the story of Noah describes the first. I believe that Rama is a cosmic Ark, sent here to save – those who are worthy of salvation.'

Similarly, the UFO flap that began in 1947 was drawn into the Heaven's Gate doctrine.

'It is a fact of record,' declared their web site, 'that a number of space alien groups or "races" have related to humans as recently as the past few decades for various reasons. These ranged from "deals" of mutual benefit – trading spacecraft technology for uninterrupted genetic experimentation – to missions of "spiritual enlightenment". All of these activities are far beneath Next Level interests . . .'

This attitude is quite common among the more transcendentally minded alien believers. They dismiss the vulgar mess of disputed sightings, contacts and abductions as being the work of lower species of aliens who are simply getting in the way of the grand cosmic project to which only true adherents have access. The cosmos is full of troublemakers.

But, though they fit neatly into postwar ufology, the real significance of Heaven's Gate is that they also link modern alien lore to the ancients. For their creed is quite clearly what, in theological terms, is known as Gnosticism.

Gnosticism is not easily pinned down. It has many forms. And it is not, as is often assumed, simply a Christian heresy. Recent scholarship suggests it has Babylonian and Egyptian roots. I would go further and suggest that it is, like belief in God, an idea that seems hard-wired into the human mind. The central creed of Gnosticism is something everybody feels to be true from time to time.

That creed is that salvation is not to be attained by faith or good works but by knowledge. As the *Catholic Encyclopaedia* puts it, 'it is markedly peculiar to Gnosticism that it places the salvation of the soul merely in the possession of a quasi-intuitive knowledge of the mysteries of the universe and of magic formulae indicative of that knowledge. Gnostics were "people who knew", and their knowledge at once constituted them a superior class of beings, whose present and future status was essentially different from that of those who, for whatever reason, did not know.'

Gnosticism sometimes feels true to us all because of the very familiar sensation that there is something we are not being told. It drives the alien cover-up conspiracy theorists. And it finds very exact fictional expression in the TV series *The X-Files*. The motto of the series was 'The Truth Is Out There'. But the truth was also concealed by government and alien adepts. FBI agent Fox Mulder will not accept this concealment and he believes in the secret knowledge that will unlock the truth and save him. He is a Gnostic, a romantic, divinely inspired heretic. He cannot rest until all the X-Files are opened and their demonic keeper, the Smoking Man, is destroyed. Gnosticism is the faith of contemporary paranoia.

It is also the faith of Applewhite and James Jones. The small group of the elect looking forward to a brighter future, the secret wisdom, the magic formulae are all Gnostic in style. Indeed, Gnosticism has always been associated with magic, an association that many orthodox Christians regard as its original sin.

Moreover, like the faiths of Applewhite and Jones, Gnosticism inclines towards the renunciation of the things of this world. Gnostics tend to be deeply pessimistic about the future of life on Earth. They see our bodies and the whole realm of matter as a deterioration of the spirit, a gross fallen condition from which the few will escape. The goal of all being is to escape this condition and 'return to the Parent-Spirit, which return they held to be inaugurated and facilitated by the appearance of some God-sent Saviour'.

Because of their pessimism, their utter rejection of the things of this world and their belief in magic and arcane wisdom that will benefit only the few, Gnostics have always been anathematized. St Paul is generally taken to be attacking the Gnostics when, in the First Epistle to Timothy, he speaks of 'profane novelties of words and oppositions of knowledge'. And, though clearly far older than Christianity, it is most often seen as a specifically Christian heresy precisely because it was so vehemently rejected by the early Christians.

There are, therefore, four key elements: secret wisdom and magic, the small band of the elect, the rejection of the body and the rejection of and by the world. The hounded, peripatetic bands led by Applewhite and Jones were modern Gnostics in all but name.

It was Applewhite in particular who deployed aliens as the agents of the Gnostic vision. In doing so, he took the extraterrestrials back into the past and raised some central questions about the postwar flap. Was this really new or had there always been these visitors? Were the aliens just contemporary versions of the demons, goblins and angels of the past? Furthermore, as Gnosticism seems to be a permanent human faith to which, from time to time, we are all drawn, the aliens that spring so easily into the minds of these heretics would appear to be our perpetual companions. Real or *real*, it looks as though their presence among us is an eternal fact of human experience.

8

Aliens and Angels

The Muons from the planet Myton were expected to land in 2001 on the 100 anniversary of the birth of the visionary Ruth Norman. Huge saucer-like ships would descend on sixty-seven acres of land in Southern California. Some of these ships would be five miles across and they would bring 33,000 Muon scientists to Earth. These beings would solve all our economic and technological problems.

Norman is the charismatic leader of a flying saucer group known as Unarius (short for Universal Articulate Inter-dimensional Understanding of Science), based in El Cajon, near San Diego. The present tense is correct because, though she died in 1993, she lives on as the Archangel Uriel. With her husband, Ernest (Archangel Raphiel), also dead, she continues to pursue her mission 'to overcome the inertia of individuals and humankind as a whole by presenting knowledge of the Interplanetary Confederation – 33 worlds joined in harmony and brotherhood'. In past lives Ruth was Mary of Bethany, Buddha, Mona Lisa, Socrates, Benjamin Franklin, Henry VIII, King Arthur, Charlemagne, Confucius, Elizabeth I, Peter the Great, Johannes Kepler, King Poseid of Atlantis and others. Ernest had been Jesus, Osiris and Satan. Charles Spiegel, a later Unarius director, had been Napoleon and Pontius Pilate.

Unarius is said by cult-watchers to be 'benign', as opposed to Heaven's Gate, which was also, probably coincidentally, based in San Diego at the time of its mass suicide. There are many such systems built around the idea of alien contact. Unarius is typical in its prophetic style, its eclecticism, its debt to the occult tradition and its offer of salvation. The long lists of the Normans' previous incarnations is also typical in its aspiration to include all of human history within the explanatory system of the creed. This historical inclusiveness is intended to show that this vision is not, like the rash of flying-saucer sightings since 1947, a new thing. It is, in fact, a very ancient thing indeed.

There are a number of tangled threads that join UFOs and aliens to the distant past. The nuts and bolts explanation of these connections would be that aliens have, indeed, always been involved in one way or another with life on Earth. Psychosocially, the eerie presence of strangers, the existence of a cosmic realm of life and the comparison, favourable or unfavourable, of that life with our own may be said to spring from the same roots as the great religions. They are all ways of explaining the strangeness and ambiguity of human experience, ways that may well be hard-wired into us by our evolutionary descent. But it is neither nuts and bolts nor psychosociology that is the issue here, it is the third realm that dominates this chapter, the belief that these beings are of neither mind nor matter, that they inhabit another place altogether.

But first, there is a short, psychosocial step back into the past to be taken. In his book *The Air Loom Gang* Mike Jay tells the story of James Tilly Matthews. Matthews, incarcerated in the Royal Bethlehem Hospital (Bedlam) in London at the time of the French Revolution, was the only man who knew of a secret gang then operating in Paris. This gang had a machine – the air loom – which could control human minds at a distance by the deployment of animal magnetism. This was an alternative term at the time for hypnosis or mesmerism; it was a term that implied a physical rather than a mental explanation for this phenomenon. Matthews believed that government and the military were being controlled by this machine and the gang was hell-bent on driving Britain and France to war.

Even in the 1790s, such an idea was not, in outline, new. Being possessed by demons that control one's mind or being in mental contact with others at a great distance are commonplaces of human experience. What was new, however, was the imagery. Matthews did not see demons, he saw political conspirators. And the device employed was not magical or Satanic, rather it was a machine, a product of human technology. The timing is important. The Industrial Revolution was under way. The world was being transformed not by gods but by human beings who aspired to be gods. Yet, in spite of this revolution of practical reason, humans remained incorrigible. They still had the same old experience of that world, an experience that involved demonic entities and the possession of minds. And so they explained those experiences in the terms that had become available at the time – the terms, for Matthews, of industrial technology.

Jay follows the imagery of the air loom and mind control down to the present day. Notably there were the brainwashing fears of the Cold War, as expressed in *The Manchurian Candidate*. Again there was the fear of the mind-controlling demon, in this case the Chinese or the Soviets. But, equally notably, there was the telepathic control of the aliens.

'Many of the earliest stories from flying saucer contactees,' writes Jay, 'the founding myths of the mass-market UFO phenomenon, revolve around alien influencing machines, the Air Loom now in orbit. We hear of brainwashing, reprogramming of minds via cosmic rays, alien wires hooked into subjects' heads to capture their thoughts, magnetic implants identical to the one Matthews located in his own brain shortly after his admission to Bedlam.'

In the fifties, the Cold War and fear of our technological hubris brought the theme into even sharper focus. It is vividly present, for example, in Evelyn Waugh's autobiographical novel *The Ordeal of Gilbert Pinfold*, published in 1957. The hero undergoes a breakdown in which he becomes convinced that a sinister man from the BBC, named, significantly enough, Angel, possesses a box that can see into and influence the minds of people. It was developed by the Russians and deployed by the existentialists in Paris to psychoanalyse people. Pinfold recovers, but he is unable entirely to dismiss the experience.

'No sound troubled him from that other half-world into which he had stumbled but there was nothing dreamlike about his memories. They remained undiminished and unobscured, as sharp and hard as any event of his waking life.'

Pinfold had glimpsed another world, invisible to this one, and, though it quite clearly seems to have been created by his breakdown, its harsh reality stays with him. Merely getting better does not provide absolute assurance that this other place does not exist. One day, therefore, he may return. Such a breakdown is not new, but the terms are. The ideas of the mechanical control of the mind and of a parallel world were in the air in the fifties, a time of technical ingenuity and paranoia.

Jay's book effectively identifies a moment when a human experience that should be rejected by modernity is, in fact, dragged kicking and screaming into the modern world. The eighteenth-century Enlightenment had aspired to clear the heavens above of their clouds and the Earth beneath of its shadows. There would be no more dragons at the edge of the map, no more demons beneath the ground, no more flights of angels in the sky and no more fairies. It is important to be aware of what a radical idea this was. As Patrick Harpur observes, for example, by the second century AD virtually everybody in the world believed in fairies, known by a variety of names. The secular mind takes it for granted that such entities are mytho-logical, perhaps metaphorical. But secular minds represent a very small proportion of all the human minds there have ever been.

Nevertheless, the Enlightenment faith was that all would be exposed to the rational gaze in the clear light of day. 'Nature and Nature's laws lay hid in night,' wrote Alexander Pope. 'God said, "Let Newton be!" and all was light.' But, as Matthews' story shows, it wasn't. The darkness had returned clad in the lineaments of the machine or, latterly, the alien. Or perhaps, as the Unarians and Heaven's Gate clearly believe, it was not darkness at all, but a much brighter light than anything dreamed of by the eighteenth-century rationalists.

The hard psychosocial line on this would be that crazy people see crazy things and they derive the details of those things from other crazy people. Gilbert Pinfold would seem to come into this category, but for the 'undiminished and unobscured' memories he retains when he is sane. Two points render the simple crazy-people hypothesis inadequate. First, the consistency of the visions over time and across cultures must say something more about the human mind than simply that it is prone to madness. As Jay says of Matthews, you no longer have to be mad to see the machine he saw, it is 'starting to come into focus for the rest of us'. Secondly, most contactees and abductees are plainly not crazy, any more than were those who saw demons and angels in the ages when people assumed we were surrounded by such beings, or who felt the controlling forces of the influencing engine in an age when they were confident we weren't. The softer psychosocial line would, therefore, be that the human mind, in all its degrees of sanity, is prone to visions and the details of these visions are certainly culturally significant and may well provide clues to the fundamental workings of the human mind. This was the insight captured by Ralph Waldo Emerson in a celebrated sentence that represents a sharp rebuke to the pretensions of the Enlightenment: 'Dreams and beasts are the two keys by which we are to find the secrets of our nature.'

Dreams and beasts, angels and demons have not been dismissed by the Enlightenment, though they may, in response to its influence, have changed their clothes. Now they may have become technologically advanced aliens with their unimaginably powerful machines. Patrick Harpur, a historian of the imagination, emphasizes the way the classic grey alien seems to have taken the place of the daimonic entities that have always stalked the world. The Sidhe are land spirits that change shape, manifesting themselves as animals or blasts of wind.

'The "greys" – or "grays" – as they have come to be known may be a new species of daimon, peculiar to the very Western culture whose orthodox world-view denies the existence of daimons. But it is more likely they are the old immortal daimons who masquerade in whatever guise suits the times. Banished from their original natural habitats, they return from outside nature, from "outer space", brandishing an "advanced technology" which duplicates the supernatural power of the Sidhe.'

This means, of course, that the aliens are not new, simply the foot soldiers of a post-1947 invasion of Earth. Rather they are very old, the parallel worlders that have haunted us since the dawn of man. Certainly, it has been noticed by many that the alien lore of the postwar world has adopted some very ancient and familiar patterns.

'Today UFO beliefs crystallize,' writes folklorist Thomas Bullard, 'around such age-old mythic motifs as otherworldly visitation, diminutive supernatural beings, the end of the world, and cosmic salvation. Even the archaic pattern of initiation and elements of fairy folklore reappear among abduction reports. UFO tales look to a distant past even as they bridge the gulf between long ago and things to come, expressing in technological idiom certain religious needs and supernatural themes otherwise lost in a secular age.'

Marshall Applewhite had between two and four thousand years of Gnostic history behind him and the mind controlling demons that haunted Matthews probably had ancestors present at the dawn of human consciousness. Seeing things in the sky or feeling the presence of the other are perpetual aspects of human experience.

There have been many specific attempts to make such historic and folkloric connections. Anthropologist Charles A. Ziegler, for example, links the Roswell story to folk narratives found in India, China, Japan, Papua, Melanesia, Tahiti and the Americas.

'The central motif of the Roswell myth is that a malevolent monster (the government) has sequestered an item essential to humankind (wisdom of a transcendental nature, ie, evidence-based knowledge that we are not alone in the universe). The culture hero (the ufologist) circumvents the monster and (by investigatory prowess) releases the essential item (wisdom) for mankind.'

This is, of course, *The X-Files* in a nutshell with Fox Mulder as the culture hero, the believer who will not be deflected from his belief, the seeker after knowledge, the dragon slayer. It is also, in variously modified forms, a structure followed by countless episodes of *Star Trek*, usually with some form of superior alien technology playing the part of the knowledge essential to humankind.

The implication of such connections is that the aliens did not suddenly arrive, unannounced, in 1947, rather they had long been part of human history – whether as literal nuts and bolts aliens, psychosocial constructs or third realmers.

The literal interpretation is, of course, that they had been coming and going for thousands of years and that, had we but eyes to see, the evidence is all around us. We live in a world partially constructed by aliens. Most commonly, artificial constructions that seem superhuman or profoundly anomalous are said to be the work of alien visitors.

Who, for example, moved the colossal megaliths on the Baalbek plateau in Lebanon? Moving such stones would be almost impossible now – so who did it thousands of years ago? What is the meaning of the inscrutable monuments at Tiahuanaco in the Bolivian Andes? Were they left by the same aliens that drew the Nazca Lines in Peru? Covering a vast area of the Nazca Plain, these were first spotted by commercial pilots in the 1920s. They are giant figures and geometrical forms, all of which have now been given resonant names like Astronaut, Man with the Hat, Executioner, Humming Bird and so on. They are so large that they can only be seen from the sky – but who was flying then? And what of the drawings on the island in Lake Tungting in China? These show people with large trunks and cylindrical objects in the sky on which more such people are standing. In the Tassili plateau of the Sahara sculpted rocks show human beings with strange, round heads, evocative, to the eager believer's eyes, of contemporary aliens. Who made the massive stone heads on Easter Island and why do they look so strange? And what does the Great Pyramid really mean? When was it built?

There are many hundreds of such examples. They are, in conventional terms, either wholly unexplained or incompletely explained. Unlike the ruins of, say, Rome or Greece, there is no written history to account for most of them. They defy our research. Moreover they are uncanny. The Easter Island heads look roughly human, but not quite. The drawings on the

Peruvian plain are almost recognizable, but, again, not quite, and why are they so vast? They defy both our reason and our knowledge. We could, we imagine, invent explanations, but they would remain inventions.

In general, such inventions would stick to our traditional modes of thought. Typically, for example, we would interpret these things as metaphorical or as illustrations of myths. But one man who decided to take them literally is Erich von Däniken, who in the late sixties, while a hotel manager in Switzerland, wrote his first book *Chariots of the Gods*. Ever since, von Däniken has energetically propagated the theory that between 10,000 and 40,000 years ago, super-intelligent aliens arrived on Earth – their first base was at Tiahuanaco – mated with primates and created *Homo sapiens*.

The creation of people, he says, 'could only have taken place by an artificial mutation of primitive man's genetic code by unknown intelligences. In that way the new men would have received their faculties suddenly – consciousness, memory, intelligence, a feeling for handicrafts and technology.' The ancient visitors made the inscrutable monuments we see around us today and the drawings, apparently of strange humanoids and spacecraft, are literal representations of their visit.

Von Däniken's great popular strength is that he is, in part, a tour guide. He talks of physical aspects of the world which we can go and see. In the age of mass tourism and travel, he attempts to re-enchant the physical world that had been disenchanted by science. These were no longer the disputed reports of mainstream ufology; these were actual, enchanted objects and they were just a plane ride away.

The discipline of investigating these things is now known as Paleo-SETI – the search for extraterrestrial intelligence in the ancient fabric of our own planet.

'Paleo-SETI research,' explains the von Däniken web site, 'is the scientific term for the revolutionary theory which maintains that extraterrestrials have visited Earth in ancient times and, in whatever form that may have been, had an influence on the development of mankind.'

It was an idea whose time – in the sixties and seventies – had come. As Keith Thompson says, people were, by then, 'tired of ufology's debate about evidence, and suspicious of science's unwillingness to take UFO reports seriously'. In von Däniken's work they found a theory which escaped both the epistemological nuances of ufology and the no-fun debunkers of hard science. It also offered a good deal more hope than the aimless broadcasting of messages in the vague direction of promising star clusters, a project that had begun in 1974. With one bound, the alien was free and he was here. Evidence of his existence littered the planet.

Von Däniken's twenty-six books have now sold more than 60 million copies worldwide. From the beginning, he was a global star. Predictably both scientists and ufologists were appalled. They scoffed at his idea of aliens who needed vast landing strips in Peru and picked holes in his history. Sumerian culture, for example, did not spring into being at the touch of an alien wand, it developed over 6000 years. But both Hynek and Vallée, seasoned observers of ufology, noted von Däniken's importance. He had touched a nerve and, said Vallée, he had dramatized 'a big credibility gap between the scientist and the public'.

He is still going strong – as you can discover at www.daniken.com. He even has his own theme park – the Mystery Park at Interlaken in Switzerland. Why not? He cannot, after all, be refuted any more than Darwin could be. Indeed, in David Brin's novel *Sundiver*, a vicious conflict rages between the Danikenites, the 'Shirts', who believe the human race was 'uplifted' by aliens, and the Darwinians, the 'Skins', who believe we did it ourselves – or, at least, evolution did it for us.

'Of one thing Jacob was sure. The Shirts and the Skins shared resentment . . . Resentment of a world in which no man any longer knew his roots for sure.'

Brin is being acute here. The fascination with von Däniken is, above all, a fascination with human origins that is not satisfied by the cold mechanism of the Darwinian account. He said that the world was strange and we were strange – not, simply, because we had failed to explain everything, but because the true explanation was intrinsically strange, it lay beyond us and our scientistic terms of reference.

In terms of alienology, von Däniken's most important

contribution was the idea that aliens are all around us, not just in space but in time. Aliens previously had tended to be creatures of the future. We would meet them when we could find them or they would come here when we were technologically ready.

(The key storyline in *Star Trek: First Contact* (Jonathan Frakes, 1996), for example, is that humans first meet aliens because passing Vulcans detect a 'warp signature' – the sign that we have developed a faster than light drive – and decide we are ready for our 'first contact'. We have attained technological maturity, without which Earth is, in the eyes of the cosmos, no more than a kind of wildlife reservation. And, of course, Klaatu comes to tell us off because we have developed rocketry and nuclear weapons.)

But von Däniken populates our distant history with aliens, immersing us in their presence. I suspect it was this idea of an alien past that inspired the first, fairy-tale line of the text prologue that opens *Star Wars* (George Lucas, 1977): 'A long time ago in a galaxy far, far away . . .'

There is a vertiginous thrill in the thought that these hurtling spacecraft are not in the distant future but in the distant past. To us *Star Wars* portrays an ancient cosmos teeming with life. Indeed, the comical variety of life forms is central to the appeal of the whole franchise. Mos Eisley spaceport – a 'wretched hive of scum and villainy' – is there to bewilder and amuse us with the sheer range of creatures on show. The cosmos is a folkloric landscape full of monsters, angels and demons. And, now, it always was.

Von Däniken, though the most effective in marketing terms, was neither the first nor the only propagator of the idea of ancient ETs. Indeed, Jacques Vallée had suggested to J. Allen Hynek in 1963 that, 'an extraterrestrial intervention might have been a factor in man's early history, specifically in the early development of civilisation and of biblical events . . . The return to such phenomena today could be explained by the need to boost our religious vacillations.'

In 1962 Josef Shklovskii, a Russian astronomer, had also speculated that ancient astronauts had visited Earth in the distant past. And, in 1956, the US Hydrographic Office discovered it possessed a medieval map that had belonged to the Turkish admiral Piri Re'is, who was beheaded in 1554. This seemed to show the east coast of South America and the coast of Antarctica, which was not discovered until 1818. It showed shoreline that should not be visible as it would be covered with ice. An anthropologist named Charles Hapgood studied this and other similar maps. They were plainly copied from earlier maps, but details like the visible coastline of Antarctica suggested they had been made in pre-Sumerian times – before the creation of writing. Hapgood concluded there had been a worldwide, seafaring civilization around 7000 BC. Others suggested the maps had been traced from photographs taken from orbiting alien spacecraft.

Aside from such theories and apparent physical evidence, it is also clear that reports of what appeared to be UFOs went back long before 1947. What, for example, are the flying machines – the Vimanas – recorded in Hindu scripture? Are they linked to the four creatures the prophet Ezekiel saw emerging from the great cloud of fire? And what were the wheels?

> And this was their appearance; they had the likeness of a man. And every one had four faces, and every one had four wings . . . Now as I beheld the living creatures, behold one wheel upon the earth by the living creatures, with his four faces. The appearance of the wheels and their work was like unto the colour of beryl; and the four had one likeness; and their appearance and their work was as it were a wheel in the middle of a wheel . . . And when the living creatures went, the wheels went by them; and when the living creatures were lifted up from the earth, the wheels were lifted up.

Alexander the Great saw a flight of UFOs – a large silver 'shield' trailed by four smaller ones – in 322 BC while he was besieging the city of Tyre. The ships blasted the city walls with beams of light and Alexander's forces marched in. Luminous discs were seen in the sky at the time of the Roman Empire, men in white clothing appeared from the heavens. Two moons shone at night. Ghost ships sailed across the sky. In 213 BC in Hadria what was described as an altar appeared in the sky, accompanied, again, by a man in white clothing. In AD 840, Agobard, Archbishop of Lyons, discovered a mob attacking three men and a woman who, it was said, had landed in a cloudship that had come from some aerial region. The four protested their innocence, saying they had been abducted by strange men who had shown them miracles. Nevertheless, the crowd would have burned them had not Agobard intervened, arguing that it could not be true that they had fallen from the sky, a line that has led to him being called the first UFO debunker. In 1493 a German scholar, Hartmann Schaeden, described a shape of fire flying through the sky; his account was accompanied by an illustration showing a cigar shape surrounded by flames. In Erfurt in 1520 a round shape with a rotating light accompanied by two fiery suns was seen.

Such accounts are repeatedly presented to suggest that the official version of the cosmos is simply wrong. Far from being the empty, lifeless place we have been taught, it has always been populated with alien beings. We, the supposedly enlightened ones, have lost sight of this. More 'primitive' people have held on to this ancient wisdom. The Dogon tribe in northern Mali believe fish gods called the Nommo brought civilization to Earth from Sirius 3000 years ago. Dogon tradition says that Sirius has a companion star. It moves in an elliptical fifty-year orbit around Sirius. This was later established as being accurate by Western astronomers. The Dogon also knew that the moon was dry and dead and that Saturn was surrounded by a ring. This is not visible to the naked eye. They also knew the planets revolved round the sun and that Jupiter had moons. Plainly the Dogon had always known something we didn't. Like the Australian aborigines in *The Right Stuff*, they could contact the cosmos without the need for our cumbersome instruments and violent rocketry.

Perhaps the world visible to pre-Enlightenment eyes was more in accord with reality than the one we had been taught to see. It was an idea that appealed to the dissident culture of the sixties. As American technology was sucked ever deeper into the mire of Vietnam, alternative accounts of the world grew ever more attractive. Indeed, *Apocalypse Now* (Francis Ford Coppola,

1979) showed a rogue colonel in Vietnam resorting to an irrational barbarism in the face of the cold ideological purity of the Viet Cong. Our warmly liberal, technocratic world had nothing sufficiently cold, fierce and brutal to confront the threat. The answer was a savage atavism of painted bodies and blood sacrifice. On the opposing side, the communism of the Viet Cong was, of course, a product of the Western Enlightenment.

In the same context, ancient aliens with superior technology or magical, godlike power became items in the hippie image banks or belief systems. The Age of Aquarius – the New Age – was one in which we would rediscover our roots not just on Earth but in the cosmos. The official account was as wrong and oppressive as sending your young men to kill 'Charley' in the jungles of South East Asia. We were ready to rejoin the fellowship of the stars.

The central insight was that we were in a fallen state. This truth had been an aspect of ufology from the beginning – the aliens were here because of our wicked nuclear experiments – but its spiritual significance was initiated slightly later, on 23 May 1951, in fact. It was then that Orfeo Angelucci, then working as a mechanic at Lockheed in Burbank, California, saw a UFO as he was driving home from work. He heard voices.

'We see the individuals of Earth as each one really is, Orfeo, and not as perceived by the limited senses of man. The people of your planet have been under observation for centuries, but have only recently been re-surveyed. Every point of progress in your society is registered with us. We know you as you do not know yourselves.'

On 23 July 1952 Angelucci was taken on board a spaceship. He looked down on Earth and a voice said, 'Weep, Orfeo . . . we weep with you for Earth and her children. For all its apparent beauty Earth is a purgatorial world among the planets evolving intelligent life. Hate, selfishness and cruelty rise from many parts of it like a dark mist.' Humanity was not aware of the true mystery of its being.

Angelucci's book *The Secret of the Saucers* was published in 1955. It would have just been one of many such works around at the time, were it not for the fact that it fell into the hands of Carl Gustav Jung as he was completing his own *Flying Saucers: A*

Modern Myth of Things Seen in the Sky. He included an account of Angelucci's experiences in an epilogue. Jung's view was not that the experiences were 'a concrete happening' but that they confirmed his own theory that they were psychic projections of newly realized archetypes of the psyche.

'Orfeo's book,' he wrote, 'is an essentially naïve production which for that very reason reveals all the more clearly the unconscious background of the UFO phenomenon and therefore comes like a gift to the psychologist.'

Jung's interest elevated Angelucci to the level of an early prophet of UFOs and aliens as spiritual messengers. His view of what Angelucci reported is not in itself sensational. It is a refined psychosocial interpretation of aliens as ancient, just as von Däniken's is a nuts and bolts version – they really came, they really built these things. But it is the third realm that is the most active in the area of aliens and the spirit and it is the third realm interpretation which Angelucci effectively ushered in.

The world was ready, as it always has been, for such an explanation. As, for example, the popular imagination began to grasp some of the strange concepts emerging from twentieth-century physics, people began to feel that perhaps the world had not been disenchanted after all. Quantum theory, in particular, with a description of matter that defied all common sense, seemed to refute the cold, mechanistic view of Enlightenment science. At the sub-atomic level, the quantum world consists of demonic, shape-shifting entities embarked upon some defiant project of deliberate confusion, of the obfuscation of common sense. Patrick Harpur goes so far as to say that the phrase 'sub-atomic particle' and the word 'daimon' are, in all contexts, interchangeable.

Furthermore, there is the resonant 'many worlds' inter-pretation of quantum theory. One way of making sense of the capricious maze we have uncovered at the heart of matter is to accept the possibility that many universes co-exist side by side, each a bifurcation from a previous one and each utterly unable to contact any other. The spectre of a third realm – or, possibly, millions of realms – haunts even the scientific imagination.

It has always haunted conventional religion. The Hindus had

no problems with their Vimanas and the Buddhist Dalai Lama seems happy with the idea of alien visitors to Earth. Indeed, the Tibetan Tulpa is a being that is brought into existence solely by the act of imagination. Tulpas can be anything you imagine, but, unlike Jung's psychic projection, they do become real and they seem to imply the existence of a realm outside ourselves and the world. They combine all three categories of alien explanation.

And the third realm is, in fact, implicit in conventional Christianity. Few Christians now believe in a heaven literally above and a hell literally below. But they may believe in the existence of such things beyond the sphere of our normal perception or imagination. In addition, the peace promised in Heaven is the peace 'that passeth all understanding' and, as St Paul says, in our earthly lives, we see only 'through a glass, darkly'. That we know that there is a place, a parallel world, and an experience of which we can, in our present condition, know nothing is the primary clue to the existence of the third realm.

In this context, the Christian visionary, who sees aspects of this other realm, can be seen as a close cousin of the alien contactee. One visionary incident, in particular, is accepted in the catalogue of alien lore. This is the incident at Fatima in Portugal on 13 May 1917. Three children saw a woman in white robes. She asked them to return to the same spot every month for six months. With each successive return there were more spectacular incidents – explosions, clouds in trees and the children were shown a vision of hell. More and more people arrived to watch. The children were warned that a second world war would follow the Great War then in progress if humanity did not improve itself. In September 30,000 people saw a globe of light moving towards them. It came down in a zig-zag pattern, the 'falling leaf trajectory' with which UFOs are frequently seen to descend. Then it rose and disappeared into the sun. In October 70,000 people saw a revolving, multi-coloured disc in the sky which plunged downwards and then, again, upwards into the sun.

Remove the Christian context and these could be mainstream UFO incidents. The white-robed figure and the glowing discs certainly echoed ancient reports now adopted by ufology. But does this endorse the Christian interpretation of Fatima? Does it not rather refute it by relativizing it as one possible interpretation among many of a phenomenon that is, historically, relatively common?

In practice, alien studies do, indeed, subject Christianity to this relativizing process. Orfeo Angelucci was told that Christ was only 'allegorically' the son of God. In fact, he was 'the Lord of the Flame', not of earthly origin but 'an infinite entity of the sun'.

'As the Sun spirit who sacrificed Himself for the children of woe he has become a part of the oversoul of mankind and the world spirit. In this he differs from all other cosmic teachers.'

But, plainly, He is only one cosmic teacher among many. The same idea is present in the communications from the committee of intergalactic beings known as the Nine. 'Tom', their spokesman, reveals that Jesus is only one of a number of perfect communicators, others being Socrates, Nostradamus, Buddha and Leonardo da Vinci. And the Pleiadians, who channel their communications through Barbara Marciniak, say the crucifixion was a 'holographic insert' into our reality.

Mainstream Christianity would be very uneasy with such revelations. And, indeed, there has always been a theological problem with the idea of alien life. Though, after Galileo, the church may have abandoned its adherence to geocentrism – the belief that the Earth was the centre of the universe – it cannot abandon its conviction that there is something unique about the human species. Why, if we are not unique, did God send his son to die a horrible death to save us? And, if there are many intelligent species on many planets, did Christ have to die the same death on every one? The thought is shocking, barbaric, absurd. The priest in Stanislaw Lem's novel *Fiasco* summarizes the problems involved in transporting the Gospel to other inhabited planets.

The image, for example, of innumerable planets with a multitude of apples where there were no apple trees, or figs that the Son of God could not curse because no fig trees grew there. You had an army of Pilates washing their hands in billions of vessels; a forest of crucifixions; crowds of Judases; and immaculate conceptions of beings whose reproductive physiology provided no room for the idea, since they multiplied without copulating. In short, the multiplication of the Gospel by all the arms of all the spiral galaxies turned our Credo into a caricature, a parody of religion. Thanks to these arithmetical jokes, the Church lost many of her faithful.

On the other hand, perhaps the aliens didn't acquire, as we did, original sin, in which case they would not need salvation. Or, even if they did fall, God is free to do what He likes, as George V. Coyne of the Vatican Observatory points out, so He may have left them in their fallen state or chosen some other means of salvation. But, as Coyne admits, we cannot know. 'While searching for rational explanations,' he writes, 'of how to confront the possibility of ETs with certain religious beliefs, we must remain open to the religious experience of God's mysterious ways.'

The predicament of religion when confronted with extraterrestrial life is captured in James Blish's novel *A Case of Conscience*, first published in 1958, in which the planet Lithia is inhabited by intelligent twelve-foot lizards. They have advanced technology, though very different from that of Earth, and they live in a condition of eerie tranquillity: 'The Lithians had no crime, no newspapers, no house-to-house communications systems, no arts that could be differentiated clearly from their crafts, no political parties, no public amusements, no nations, no games, no religions, no sports, no cults, no celebrations.'

Lithia seems to be an improbably secular utopia. A Roman Catholic priest who arrives with the astronauts from Earth is troubled by this. The Lithians are, he observes, 'a Christian people lacking nothing but the specific proper names and the symbolic appurtenances of Christianity'. And then this priest is struck by heresy. This is, he realizes, 'a planet and people propped up by the Ultimate Enemy'. The Devil has created Lithia to demonstrate to humans that their religion is pointless. Paradise can be had in this life without any of its mumbo-jumbo. Lithia is an attempt to destroy the faith of humans. This is heretical since it is a form of Manichaeism, the belief in an eternal war between good and evil in which the forces of evil are creative. In orthodox Catholic theology, they are not, they require human action to be unleashed. Finally, the priest is driven to exorcize Lithia and, at that moment, it is destroyed, though it is unclear whether this was, in fact, the result of a dangerous experiment with the planet then being conducted by human scientists.

In spite of the problems, many Christians have, one way or another, admitted aliens into their theologies. Some fundamentalists believe UFOs bring demons to the world. This idea finds some support in Thomas Bullard's observation that the intrusive, piercing instruments used by aliens on humans in the course of abductions are similar to the forks wielded by devils in Christian art. He, in turn, finds backing from Keith Thompson, author of *Angels and Aliens*: 'Over forty years of UFO sightings have led to the creation of an alien hierarchy (tall, Nordic-looking "blonds" and short, repulsive "grays" being only two types) no less daunting than the multileveled choir of angels, a richness bound to be documented in vivid detail some day by some enterprising folklorist.'

In practice, however, the third realm of the aliens has tended to defy conventional religious categories. Indeed, Patrick Harpur sees Christianity as the original problem, prior to the Enlightenment: '. . . it is a purely Western peculiarity to confuse the literal and the physical. It is the result of the Christian polarizing of soul and body. Outside Christendom, and other monotheistic religions, the soul is as quasi-material as the body is quasi-spiritual – both forming a daimonic whole. We are fluid organisms, passing easily between this world and the other, between life and death.'

Most typically the third realm seems to consist not of a religious singularity, but of a teeming plurality of entities subject to an equally teeming plurality of desires and demands.

There is an important connection here between the teeming third realm and the teeming world of computer games. Video games began to appear in the late seventies. Prior to their appearance, a popular board game named Dungeons and Dragons had established a fad for magical, potentially third realm imagery. Video games began as electronic ping-pong, but, as processing power increased, it became possible to render detailed and complex imagery. This imagery was frequently routinely militaristic. But, from the beginning, an alternative culture established itself on the screen. This began as Space Invaders and Pac Man and moved on to exotic creations like Sonic the Hedgehog, entities that seemed to emerge from a realm behind the screen and whose behaviour was entirely determined by the internal logic of that world. Eventually, full-blooded realms of demons and goblins became the dominant imagery in the games. There were, in addition, oddities like the massively successful Myst. This was set in a hybrid landscape that seemed both ancient and modern. The game was not so much about solving a puzzle as inhabiting one. In Myst, the computer realm had become autonomous, a world neither of our mind nor of matter. That this was, indeed, the third realm was made explicit by the publicity surrounding the launch of Sony's PlayStation 2 games console. This, it was said, was 'the third place'.

Computer games are about absorption in another logic that is neither that of the player nor that of the ordinary world. Such evidence, on top of the new worlds of alien lore, suggests that the human mind is naturally drawn to a riotously plural otherworld, one in which the beings and laws of this one do not exist and yet one which bears a tantalizing, tangential relation to the world we know. This human tendency towards an alternative plurality caused Calvin to despair – 'Surely, just as waters boil up from a vast full spring, so does an immense crowd of gods flow forth from the human mind.' The lure of the third realm offended the Protestant reformer in particular because it is undisciplined and thwarts the efforts of strict religion to hold the human imagination in check. But that is the way it is in the third realm. It is, however, not entirely undisciplined for it remains a place where we must still learn to discern good from evil.

'The majority of extraterrestrials are here for your upliftment, though there are also those who are here for other reasons,' the Pleiadians told Barbara Marciniak.

Again this is a reference back in time. The multiplicity of aliens equates to the multiplicity of gods, demons, goblins, elves and fairies that once accompanied – for good or ill – human life. As

Whitley Strieber points out, even ideas of hybridization with third realmers are nothing new.

'The Roman historian Suetonius maintained that Caesar Augustus was the product of relations between his mother and an incubus. Plato was also believed to be the issue of some sort of peculiar coupling, as was Merlin the magician, born of an incubus and one of the daughters of Charlemagne.'

The key, though perhaps paradoxical, point here is that, as UFO sightings blossomed into contact stories and then increasingly florid tales of abduction, the third realm began to seem like the most rational explanation. Between 1964 and 1968 there was a sightings pandemic and, of course, abductions got under way with the revelation of the Betty and Barney Hill case. As Thomas Bullard observes, the whole phenomenon seemed deliberately designed to defy rational analysis.

> Respected investigators like John Keel and Jacques Vallée declared the phenomenon elusive by nature and pro-nounced that UFOs were not just stranger than anyone imagined but stranger than anyone *could* imagine. They disappeared, materialized and dematerialized, shape-shifted, and fraternized with monsters or other paranormal events. The occupants acted more like fairies or spirits than scientific explorers. UFO research came to mean specula-tion about alternative universes or an impending shift of human consciousness as nuts-and-bolts machines dissolved into a wooly New Age spirituality of the counterculture era.

After the first bifurcation of ufology caused by the abduction phenomenon, here was another fork in the road. The aliens had been concerned scientists, then they were invaders, now they were daimonic entities, playing with our first- and second-realm sensibilities. *Star Trek*, ever alert to such developments, included one such entity in its *Next Generation* series. Q is the mischievous, extradimensional being who periodically appears to tease Captain Picard and the crew of *Enterprise* with his dandyish contempt for our merely human perceptions of the cosmos. Douglas Adams also picked up on the idea of aliens as compulsive tricksters in his *The Hitchhiker's Guide to the Galaxy*. Here the alien Ford Prefect explains to the human Arthur Dent how he arrived on Earth – with the aid of a teaser.

‘ "A teaser? Teasers are usually rich kids with nothing to do. They cruise around looking for planets which haven't made interstellar contact yet and buzz them."

‘ "Buzz them?" Arthur began to feel that Ford was enjoying making life difficult for him.

‘ "Yeah," said Ford, "they buzz them. They find some isolated spot with very few people around, then land right by some poor soul whom no one's ever going to believe and then strut up and down in front of him wearing silly antennae on their heads and making beep-beep noises. Rather childish really." ’

The capricious nature of the demonstration and 'the poor soul whom no one's ever going to believe' neatly captured the predicament of the ufologists. The aliens did crazy things and they did them in front of unreliable witnesses. And this wasn't just in fiction. In 1977 UFOs buzzed Ripperston Farm in England, an alien peered in through the window, terrorizing the Coombs family. But all these visitors did was repeatedly to release the Coombs' herd of cattle from their field, once instantaneously relocating the presumably bewildered cows to a farm nearby. In 1967 in Chitterne, near the UFO hotspot of Warminster, a herd of cows vanished completely, to be returned to their field the next day.

These things were crazy. They were all, therefore, nonsense or they were a manifestation of another logic, a third-realm logic. If the latter, then there can be no point in even trying to embrace the aliens with human rationality. But they are, nonetheless, here and so either we must have got it wrong or we have not yet been uplifted to the necessary higher level of understanding.

This is, in general, the position of groups which have contacted aliens through occult means such as mediums and channels. The Pleiadians who channel through Barbara Marciniak, for example, tell us we are about to undergo 'a dimensional shift that will lessen the density of the third dimen-sion so that you will move into higher dimensions in which the body does not have such a solid state'.

In fact, 500,000 years ago Earth had a highly evolved human civilization. Earth was supposed to become a Living Library, a centre of information for the galaxy. But there were disputes surrounding the project and, 300,000 years ago, certain 'creator gods' raided the Earth. These new rulers of the planet are the beings spoken of in the Bible and in Babylonian and Sumerian tablets. They wanted to keep humans in ignorance so they rearranged our DNA. Originally we had twelve-stranded DNA molecules, but the creator gods reduced the number to two – the double helix we now know so well.

Through Marciniak we are told: 'Anything that was unnecessary for survival and that would keep you informed was unplugged, leaving you with only a double helix that would lock you into controllable, operable frequencies.'

All that we have left is 'the frequency of the orgasmic experience in sexuality so that you could remember your higher identity'.

Many entities feed off consciousness and human emotions, in particular, are food for others. 'When you are controlled to bring about havoc and frenzy,' say the Pleiadians, 'you are creating a vibrational frequency that supports the existence of these others because that is how they are nourished.'

But there are greater beings who live off the vibration of love and who now wish to 'reestablish the food of love on this planet'. We are to be raised up to our original condition, a project that 'will affect the entire universe'.

Earth has similarly central significance for the Nine. They say the planet is stuck and its evolutionary immobility is holding back the evolution of the entire universe. Through their medium, Phyllis V. Schlemmer, Tom, the spokesman for these intergalactic beings, tells us: 'Man is now coming out of the true dark ages of the planet and becoming aware of the existence of other life forms in other parts of the universe. Men have always assumed there was something sitting up there taking care of their problems, but they also assumed through their ego that they were the only existence and that this being called God was only concerned with them. Man has now to understand that there are other forms of life and that the universe does not revolve, or evolve, just around man.'

The Nine agree with the Pleiadians that Earth is the only planet on which there is free will and individual choice. On all other planets consciousness is collective. The Nine also give a fantastically elaborate rewriting of Earth history involving multiple previous civilizations that were created by extraterrestrial beings – Hoova, Altea, Ashan, Ancore, Aragon – whose descendants now form the various human races. Africans are different, however; they are the original beings that evolved on planet Earth.

Tom confirms that alien ships and beings have been captured by governments in the United States and Switzerland. Gene Roddenberry, creator of *Star Trek* and known to its fans as 'the great bird of the galaxy', wanted a little more than this. For a time he attended Schlemmer's sessions and *Deep Space Nine*, one of the *Trek* incarnations, is said to be named after Tom's committee. He asked why the Nine didn't give definite evidence of their existence.

'It is of great importance for you to understand,' replies Tom, 'that the governments of your Earth have refused to believe, or to convey to the people, our existence. If there were an attempt by the civilizations to land upon Planet Earth in a mass situation, which in truth will come to pass in the course of time, the people on Planet Earth would panic, for they have not the understanding, the knowledge, that we would mean no harm to them.'

The third realm is what we have most assiduously covered up, first with our monotheism and, secondly, with our Enlightenment scientism. It exists, though not as matter or mind does, but in a way we cannot understand because we have lost the power to do so. Humans have grown old and ignorant, they have forgotten what the world really is. 'Perhaps an angel,' John Ashbery writes, 'looks like everything/ We have forgotten . . .' We must regain the clear vision of childhood.

This brings me to one of the most important themes in alienology – the theme of the child. The idea that monotheism and scientism have blinded us parallels the familiar romantic theme that growing up has also blinded us. Childhood in the romantic era became not just a preparation for adulthood but a distinct other realm where different rules applied. The glory that had passed from the Earth in the mind of the adult William Wordsworth was the vision of natural unity he had seen in childhood. We must be reborn to discover that vision.

This idea is most spectacularly realized in the film *2001: A Space Odyssey* (Stanley Kubrick, 1968). The underlying plot of this film is similar to the story told by the Pleiadians or the Nine. Some alien intelligence is leading us up an evolutionary pathway. The signs of its presence are great monolithic blocks – evocative of those at the Baalbek plateau – that appear at key moments in history. At the climax of the film, the astronaut Dave Bowman appears to descend into a monolith. In the hallucinatory spectacle that follows, he grows very old and seems on the point of death when the monolith makes its final appearance. Again he seems to merge with the object, only to re-emerge, gazing down with infinite gentleness at Earth, as a foetus, ready to be reborn.

The quest of Roy Neary in *Close Encounters of the Third Kind* is a quest for the clear vision of childhood, away from the adult clutter of family life. It is also, of course, in Charles Ziegler's terms a case of the culture hero subverting the monster in pursuit of wisdom essential to mankind. 'Monsieur Neary, what do you want?' asks Claude Lacombe, the movie's Jacques Vallée figure. 'I just want to know that it's really happening,' replies Neary. The question requires a bigger answer than it seems to get. But, in fact, that 'just' is ironic. If it is really happening, then Roy Neary is being reborn as surely as Dave Bowman.

In the movie a golden-haired child is also snatched by the aliens. Retrieving him becomes the goal for his mother, Neary's partner in his quest. When the aliens arrive at this child's house, the first sign that they are there is a strange ballet of all his electrical toys, booted into life by, presumably, the electro-magnetic field generated by the alien technology. All of this is, of course, leading up to the appearance of the aliens themselves. This comes after a fantastic display first of their smaller ships and then of their colossal, spiked, turreted, glowing mother ship, its structure evoking an American city seen at night reflected in water. It all suggests vast, heroic and very adult engineering, but, when the aliens emerge, they are like naked children – say, five-year-olds. They place their childish hands on Neary, gently, tentatively. They seem as childishly thrilled by the encounter as he does.

With *E.T. The Extra-Terrestrial* (Steven Spielberg, 1982) Spielberg abandoned mere suggestion in favour of the direct statement that the coming of the alien is a childhood event. E.T. himself is as small as a child and has a child's large head and big eyes. At one point, he is concealed from the adults in a pile of cuddly toys. The culture hero this time is the boy Elliott and the monster is the authorities, the Men in Black who would steal E.T. and subject him to medical examination. But the wisdom that is at stake is not just the existence of the extraterrestrial, it is also the wisdom of childhood. Elliott knows more than the adults about E.T., indeed, under examination, his physiology seems to bond with that of the creature. The triumphant announcement that E.T. has DNA – i.e. he is one of us – is only an endorsement of Elliott's intuitive perception. Elliott just *knows*, in stark contrast to the adults who must struggle with their cumbersome machinery and analytical tools for such wisdom.

This glorification of the alien as child is plainly based on the belief that there is something fundamentally wrong with the adult world. Neary's rejection of his family and the rows that constantly sweep through Elliott's household are emblematic of this dysfunction. The grim reality of broken family life – the father has gone – is contrasted with the brilliant, shining goodness of the alien. This is, in highly modified form, a continuation of Klaatu's rebuke and an echo of John Mack's concern, as well as that of the mediums and channellers that Western scientism has destroyed our capacity to see. We are corrupted, earthbound beings who need to be reborn to the cosmos through the agency of the alien. This would seem to endorse a Jungian interpretation of the aliens as projections of our anxieties about the age in which we live. We yearn for rebirth because we feel we have gone too far, grown too sophisticated and lost the eyes of childhood.

But, of course, there is an anxiety about childhood as well. In itself, it is an alien realm. *Twilight Zone: The Movie* (John Landis, Steven Spielberg, Joe Dante, George Miller, 1983) is a collective enterprise in which only Dante's segment really works precisely because it is a savage reversal of Spielberg's alien as sacred-child theme. In this a child, Anthony, is terrorizing his household. He has acquired extraordinary powers that allow him, for example, to close up his sister's mouth or create monstrous, cartoon animals. 'All I have to do is wish for something,' says Anthony, 'and it happens. I can do anything, anything.'

Dante is a director who appears to exist solely to provide the flipside of Spielberg's vision. This seems to be deliberate as Dante is a protégé of Spielberg. In *Gremlins* (1984), he showed childlike creatures as wildly destructive, cynical and sophisticated beings. And, the best joke of all, in *Explorers* (1985) human children encounter aliens who seem to be strangely frivolous creatures, the reason being, as we finally discover, that they are, in fact, alien children. Dante's childhood world is frightening for the same reason that Spielberg's is consoling – it is a place set apart from the adult world, subject to a different logic. But, for Dante, this seems to be a suffocating idea. The adults in *Twilight Zone* are trapped in Anthony's childish moods and vengefulness.

The world of the child can, in fact, be terrifyingly alien, disorientating, shocking. It was an idea vividly captured in Robert Sheckley's story 'Restricted Area'. On a mysterious planet objects move about and then stop for no apparent reason. There is a giant metal column. Eventually we discover this is a child's world. These are clockwork toys and the metal column is the key that winds them up.

On the other hand, reconciling ourselves to our children may be the whole point. In the film *K-Pax* (Iain Softley, 2001), Prot, a mental patient who insists he is from a distant planet, explains to his doctor that the physics of the cosmos will result in an eternal recapitulation. This means that the life you now lead must be morally bearable.

'I wanna tell you something, Mark,' says Prot, 'something you do not yet know, that we K-Paxians have been around long enough to have discovered. The universe will expand, then it will collapse back on itself, then will expand again. It will repeat this process for ever. What you don't yet know is that when the universe expands again, everything will be as it is now. Whatever mistakes you make this time around, you will live through on your next pass. Every mistake you make, you will live through again, and again, for ever. So my advice to you is to get it right this time around. Because this time is all you have.'

As a result of this advice, Mark is reunited with his previously alienated son. The alien restores the familial unit which physics has deemed to be eternal.

The child is thus often used to embody the whole of the alien experience – good or bad – and childhood is the correlative of the third realm, a parallel but imperfectly accessible universe. Either the alien is a child, as in Spielberg, or the child is an alien, as in Dante. Both cases are a reflection of our own anxiety about our adult severance from the experience and perceptions of childhood. We may feel love when we look at our children, but we also feel a disorientating sense of their otherness. We cannot imagine the world through their eyes because we have forgotten. In fact, forgetting is a necessary accompaniment to growing up – remembering too clearly would be a form of incapacity. But the loss is heartbreaking to the romantic imagination – 'The things that I have seen,' wrote Wordsworth, 'I now can see no more' – and it is the romantic imagination that produced Spielberg, perhaps the single most potent force in the lore of contemporary alien fiction.

To the post-romantic and postmodern imagination, the loss of contact with childhood is a reminder of the irredeemably relative nature of our understanding of the world. The fact that the child sees and inhabits a different world from our own signals the arbitrary nature of human reason. This becomes a nightmarish revelation that we are locked in, shut down by the narrowness of our view. The child is simply and horrifically other.

Spielberg may offer redemption and Dante his comic nightmares, but both visions share an immense depth of sadness. They are both about a catastrophic loss that we cannot avoid, a loss of contact with another world. Once again the alien tells us what it cannot itself stop telling us: we do not belong.

Philip Kindred Dick

Philip K. Dick died of a stroke on 2 March 1982 at the age of fifty-three. In spite of the claims of his most ardent fans, he can rest in peace, secure in the knowledge that he was not the greatest writer of his time, the Franz Kafka of the second half of the twentieth century. He was brilliant, original, dementedly imaginative and heroically paranoid, but he was a sloppy writer. Even his best books cry out for a radical rewrite. But if there is some other form of greatness that consists of an excess of sensitivity to the age in which one is born, then as a man, as a life, Dick was unquestionably great.

For Dick, as he wrote in 1980, 'The Martians are always coming.' But, sadly, they never arrived. He died three months before the release of *Blade Runner* (Ridley Scott, 1982), a film based on his short story 'Do Androids Dream of Electric Sheep?'. It has a decent claim to be among the greatest SF movies of all time and it inspired a rush of movies based on Dick stories – *Total Recall* (Paul Verhoeven, 1990), *Screamers* (Christian Duguay, 1995), *Impostor* (Gary Fleder, 2001), *Minority Report* (Steven Spielberg, 2002) and *Paycheck* (John Woo, 2003). Death denied him the phenomenal wealth his ingenuity was to earn. Minority Report grossed $132 million, Dick was paid $130 for the original story in 1956. He sold 'Paycheck' for $195 in 1954; Paramount paid $2 million for film rights in 2003.

Dick was a beatnik, hippie cyberpunk, a drifter. He was born on 16 December 1928 in Chicago. His twin sister, Jane Charlotte, died at six weeks after his mother, Dorothy, scalded her with a hot water bottle. 'She died of neglect and starvation, injury, neglect and starvation,' he said.

'Out of his loss,' speculated Richard Corliss in *Time* magazine (14 January 2004), 'Phil was granted an invisible playmate and eternal soul-mate – also, perhaps, someone against whose impossible ideal all other women, from his mother to his five wives to the cliché-ridden females in so many of his stories, would be measured and found wanting.'

His parents divorced and the family moved to California. He studied German at college. But the family was poor and he supplemented their income by selling records, being a DJ and, from 1952, by selling short stories to SF magazines. In 1953 he had twenty-eight stories published and thirty-five in 1954. In 1955 he published his first novel. He didn't like SF – 'The early fans were just trolls and wackos,' he said of his readers, '. . . terribly ignorant and weird people.' He always thought the appeal of SF was 'pre-adult'.

'I have written and sold twenty-three novels,' he said in 1966, 'and all are terrible except one. But I am not sure which one.'

He was, in fact, intending to be a mainstream novelist, but the SF magazines of the time at least guaranteed him publication. Anyway, he was too good at SF. In 1963 he won the Hugo Award for *The Man in the High Castle*.

But the SF audience was specialized and relatively small. Dick was never to make much money. He had to continue to turn out novels and stories at a phenomenal rate. By the time he was thirty he had written thirteen novels and some eighty short stories. He never wrote as carefully as he should have, nor did he ever achieve a breakthrough into mainstream literature. He associated naturally with the countercultural movements of the time – the beats, the hippies and even the Communist Party – and his opposition to the Vietnam War meant there was an FBI file on his activities.

By the 1970s Dick had more or less already created the content that was to dominate subsequent SF. Cyberpunk and its leading practitioner, William Gibson, could not have existed but for Dick. His futures do not gleam, they glow darkly. They are suffused with street life, innovative drugs and uncertain realities. His robots do not know they are robots and his humans suspect they might be machines. The supremely paranoid idea that reality is an artificial construct – the idea that inspired the *Matrix* films – was brought to its first refinement by Dick.

'We live in a society,' he wrote in an essay in 1978, 'in which spurious realities are manufactured by the media, by governments, by big corporations, by religious groups, political groups. I ask, in my writing, What is real? Because unceasingly we are bombarded with pseudorealities manufactured by very sophisticated people using very sophisticated electronic mechanisms. I do not distrust their motives. I distrust their power. It is an astonishing power: that of creating whole universes, universes of the mind. I ought to know. I do the same thing.'

He had used drugs extensively himself and he had acquired a sense of the arbitrariness of the realities we choose. Modernity and the machine had intensified this reality crisis. Modernity had rendered materiality questionable. 'Things' were no longer stable and now, as machines, they were taking on a life of their own.

'The ultimate in paranoia is not when everyone is against you,' Dick said, 'but when *everything* is against you. Instead of "my boss is plotting against me", it would be "my boss's *phone* is plotting against me". Objects sometimes seem to possess a will of their own anyhow, to the normal mind; they don't do what they're supposed to do, they get in the way, they show an unnatural resistance to change.'

And yet there was a tremendous struggle, both in Dick's style and in his imagination, between the mental breakdown threatened by such visions and the urge to stabilize the world. Stylistically, he often tries to impose a Chandleresque urbanity on his stories – another of his legacies to Gibson – but he does so in the face of nightmarish spectacles that defy all such pretensions. Like the good beat he is, he tries to impose 'cool' on a meaningless world. This battle – between urbanity and catastrophe – emerges in almost everything he wrote.

'I became educated to the fact,' he wrote in 1977, 'that the greatest pain does not come zooming down from a distant planet, but from the depths of the heart. Of course, both could happen; your wife and child could leave you, and you could be sitting alone in your empty house with nothing to live for, and in addition the Martians could bore through the roof and get you.'

Meanwhile, in his imagination, he seemed to want to anchor the shifting planes of reality in some form of truth. God features a lot in Dick.

'After all,' he writes in *The Three Stigmata of Palmer Eldritch* (1964), 'the creature residing in deep space which had taken the form of Palmer Eldritch bore some relationship to God; if it was not God, as he himself had decided, then at least it was a portion of God's Creation.'

And this is from *Our Friends from Frolix 8* (1970): ' "God is dead," Nick said. "They found his carcass in 2019. Floating out in space near Alpha."

' "They found the remains of an organism advanced several thousand times over what we are," Charley said. "And it evidently could create habitable worlds and populate them with living organisms, derived from itself. But that doesn't prove it was God." '

The problem in both these cases is that God, if He exists, has been relativized. Palmer Eldritch inhabits drug-induced hallucinations as God. Is he not, therefore, as real as any previous God? And, in Frolix 8, a being that can do everything God can do is not, in fact, God. But he may still be the god we have all been worshipping. Such issues would not be issues at all if Dick were a rampant postmodernist like his successors, content with the machine realm in which truth had been abandoned as a quaint oddity. But Dick remains a modernist, paralysed in contemplation of the ruins rather than eagerly moving on. He wishes to preserve what fragments he can of the human.

'Faced with the failure of all totalizing and redemptive schemes,' writes Erik Davis in his book *TechGnosis*, 'Dick came down to nothing more than the drive to remain human in an often inhuman world. In contrast to the exhausted skepticism of the post-moderns or the juvenile glee of the posthumans, Dick never abandoned his commitment to the "authentic human", which he tentatively described as the viable and elastic being that can "bounce back, absorb, and deal with the new."'

And, in fact, Dick did finally find God. On 2 February 1974 he received a dose of Sodium Pentothal after the extraction of an impacted wisdom tooth. More painkillers – Darvon – were sent round to his house. The female courier was wearing a pendant consisting of two intersecting arcs, a design to which he gave the name 'vesicle pisces'. In fact, the phrase was his own invention. He may have meant 'vesica piscis', a symbol made of two intersecting circles in which the centre of each circle lies on the circumference of the other. This symbol became the fish sign used as a mutual recognition code by early Christians.

The visit and the symbol provoked a series of visions in Dick. He described them as laser beams and geometrical shapes with occasional glimpses of Jesus Christ and Ancient Rome. He began to lead a double life, both as himself and as a Christian persecuted by the Romans. He accepted these visions as being sent by an entity he referred to as God, Zebra or, most commonly, VALIS – Vast Active Living Intelligence System. He seemed to think this was some kind of satellite orbiting Earth and this entity used 'disinhibiting stimuli' to prepare subjects for communication. The courier's pendant had been such a stimulus.

VALIS told Dick that his son's life was in danger. Doctors said he was fine, but Dick insisted they look at him again. They discovered an inguinal hernia. There were also episodes of glossolalia – speaking in tongues.

For the remaining eight years of his life, Dick wrote novels around the theme of VALIS and also a million-word journal called the Exegesis. This was based on the theory that the Roman Empire never ended. It was the empire of materialism and, having suppressed the Gnostics, the Romans kept the world in a deluded state in which the people were crushed by their devotion to material goods. Both the KGB and the FBI were, meanwhile, plotting against him and breaking into his house. The cartoonist

Robert R. Crumb produced an illustrated version of all this entitled *The Religious Experience of Philip K. Dick*.

Dick is important, first, because he downgraded the future. Unlike previous science fiction writers, he was not interested in technological details.

'Phil's approach to technology,' Lawrence Sutin, his biographer, wrote, 'was, simply, to make up whatever gizmo he needed to keep his characters' realities in suitably extreme states.'

The future turns out to be at least as bad as the present. In fact, usually, it is worse, because human failings and crises have been amplified by technology. 'If You Find This World Bad, You Should See Some of the Others' was the title of a speech he gave in 1977. In that speech he said: 'Often people claim to remember past lives; I claim to remember a different, very different present life . . . I rather suspect that my experience is not unique; what is perhaps unique is the fact that I am willing to talk about it.'

Blade Runner may not have been a very accurate version of Dick's story, but it did capture that overpowering Dickian sense that the future is much like the present only more so – more crowded, more greedy, more violent and more baffling. There had been dystopian visions of the future before Dick, but they tended to portray a shiny, high-tech hell. Dick's hell was dirty and chaotic, the life of the street but with androids. In *Blade Runner* Ridley Scott turned this into a new stylistic orthodoxy – the future as mess. Scott repeated the effect in *Alien* (1979) in which the interior of the ship and even the alien itself, designed by the Swiss artist H. R. Giger, were suffused with an oily, warehouse sheen as if the Industrial Revolution had returned to haunt the information age.

But, secondly, Dick glimpsed the centrality of the alien in the postwar world. He was himself a stranger in a strange land, a troubled drifter. In his madness he lived in the third realm of aliens and angels. The world was alien to him and he was alien to it. He understood the eternal truth that we don't fit and he saw how modernity had heightened and dramatized our discomfort.

'The fish sign causes you to remember,' he wrote in his Exegesis, 'Remember what? . . . Your celestial origins; this has to

do with the DNA because the memory is located in the DNA . . .
You remember your real nature . . . The Gnostic Gnosis: You are
here in this world in a thrown condition, but are not *of* this
world.'

10

Alien Suggestion

Many, perhaps most, alien abduction accounts are retrieved under hypnosis. This is necessary because of the amnesia imposed by the aliens on abductees either to preserve the secrecy of their operations or to protect their human subjects from the mental disturbance such memories might provoke. As a result, abductions cannot usually be remembered in detail in normal states of consciousness. Just as Philip K. Dick could see the truth of the world only in the abnormal condition of his mental break-down, so the abductee can be made to remember his experience only in the abnormal condition of the trance. Hypnotic regression – taking people back to past events – can, it seems, break down the barrier of amnesia.

Plainly, if this imposed amnesia is entirely effective, it could mean that any one of us, possibly all of us, may have been abducted. We may be living contented, perhaps slightly bored, lives, ignorant of countless exotic and inexplicable experiences which have simply been deleted from our conscious memories. The 'real' world would then be a place of a permanent and pervasive alien presence of which we know nothing. We would be walking around in a dream which we had been made to believe was reality.

But the Roper Poll asking people about events that suggested abduction – missing time, paralysis, awareness of the presence of unseen others and so on – worked on the assumption that this deletion was imperfectly executed by the aliens. It left behind reminders that, ideally, should have been removed along with the experience. It has been suggested that this is because memories are stored holographically; they are present in every part of the brain. All the aliens can do, therefore, is block them. Hypnotism can unblock them.

That, of course, is a nuts and bolts explanation. There could be a third-realm explanation in which no amnesia effect occurs. In this case, what we see under hypnosis is a glimpse into a third realm, neither of the world nor of the human mind. Hypnosis simply tears away the bandage of convention from our eyes. Normally we see through a glass, darkly; under hypnosis we see perfectly. The aliens are inhabitants neither of our minds nor of matter, but of something else which we are unable to grasp or apprehend except under hypnosis.

Or there is the psychosocial version. The accounts narrated under hypnosis must be artefacts of the hypnosis itself. Hypnosis provides an occasion for fantasy and the details of these fantasies are derived from the stories available in the culture. If this is true, then the effects of hypnosis are quite staggeringly powerful. Thousands of people have been so convinced by what they have 'remembered' under hypnosis that it has persuaded them that their previous recollections in normal states of consciousness have been utterly false. Moreover, they have produced detailed and often very frightening narratives, the reality of which transforms their view of the world for ever.

Hypnosis – what it does, what it means – is the issue here, not only because of its use in unlocking the memories of abductees, but also because it seems to stand at the crossroads where the three conventional explanations of the alien experience meet and diverge. It is a demonstration either of the abject limitation of our everyday consciousness or of its tremendous power to hold the wayward human mind in check.

It is not, however, something that can be understood in abstraction. There is plainly a distinct experience involved that is not fully explained by phrases like 'in a trance'. So I asked Dr David Oakley of University College London to hypnotize me.

Oakley has, in the past, hypnotized people who have never claimed to be abducted by aliens. Yet, under hypnosis, they have produced compelling and detailed abduction accounts. He demonstrated this at The Skeptic Society and, for an hour and a half, the audience heard a man describe the interior of an alien ship and the removal of samples from his skin. The man did not afterwards become a believer in the reality of the experience. One member of the audience, however, did call out, 'Yeah, very interesting, but how do you know he wasn't actually remembering a real abduction experience?' It was, Oakley says, 'a chilling moment'. But, if not real, where do the experiences come from?

'Maybe they occasionally go into trances and there are little gaps where they have those experiences which they forget and then retrieve later. My take on it, though, is that it is much more likely that people actually search through unconsciously everything they know about aliens.'

In doing this, if they encounter a believer, they may well be persuaded that they are remembering rather than creating an experience. Oakley mentions the condition of torticollis, a spasm of the neck in which the head is pulled to one side. It is thought to afflict about three in every 10,000 people. Sufferers with this condition have been hypnotically regressed by therapists who believe in the use of hypnosis as a way of retrieving memories of past lives. Typically, patients discover that one of their past lives ended with them being hung. The neck spasm is a physical memory of their execution.

Oakley, not being a believer in these things, does not lead his subjects into such convictions. In fact, he does not lead them at all. It is not necessary. He finds they rapidly generate their own narratives and even reject some of his suggestions as to what happens next as being out of step with their own story.

This is a difficult process to understand from the outside and so I returned to Oakley's office in Bloomsbury to attempt to understand it from the inside. I had never previously been hypnotized and had always assumed I would be a poor subject, possibly because I would be too analytical about the process or possibly because I would simply be too perverse. In the event, neither analysis nor perversity proved a potent enough defence against Oakley's powers.

Even without hypnosis, Oakley could make me do strange things. He gave me a long string with a brass weight at the end and told me to stand up, hold it at arm's length and focus my gaze on the weight. He then told me to imagine it was the pendulum of a clock, but I was to keep it still. At the periphery of my vision, I noticed that he was swaying from side to side. He kept talking of this weight as a pendulum and then, to my amazement, it began to swing from side to side. I was aware of doing nothing whatsoever to cause this. Then I was to imagine the swing slowing down. It did so. And then to imagine a breeze was blowing towards me causing the weight to swing back and forth. Again it did so. By this time I was laughing in disbelief at my own suggestibility.

This device is known as Chevreul's Pendulum after the French chemist Michel-Eugène Chevreul who, in 1833, produced a paper on the workings of this 'magical pendulum'. What it demonstrates is an ideomotor response. If you focus on an idea in your head, you begin to make it happen whether you consciously intend to or not. I was making tiny muscular movements with my hand that caused the pendulum to move. These muscles were effectively under Oakley's rather than my control. These movements were unconscious. In fact, they would have had to be unconscious since the delicacy of the movements – particularly the smooth transition from the side-to-side to the to-and-fro swing – would have been beyond my conscious abilities. The sensation, as Oakley had warned me in advance, was of an unnerving powerlessness.

He also pointed out that, since I was unaware of doing this, I could potentially have been persuaded that the effect was magic. He could be a sorcerer, putting some kind of influence on me or the pendulum. He was, of course, doing both, but with my unconscious cooperation. The sorcery lay in accessing responses beneath my conscious control.

The hypnosis was equally startling. There was a preparatory

conversation in which I provided him with a description of a special place – the riverside garden of my house in Norfolk – in which I felt comfortable. He then put me under remarkably rapidly and, as far as I could tell, remarkably deeply. I recall him saying, 'Don't worry about logic because there isn't any logic.' And, 'Just go with it.'

He told me to imagine my left arm was attached to a helium balloon and, obediently, it floated in the air. He told me my eyes were glued shut and they were. He told me to describe my special place and I did so in infinitely more detail than I had when conscious. But there is a crucial point about this experience. He had told me to 'go' there and, indeed, in one sense, I had gone. But, in another sense, I hadn't. I could see Norfolk quite clearly but I knew I was imagining it. I could hear the traffic noise in his office and even, on a couple of occasions, the very loud sirens of ambulances or police cars passing beneath his window. If you had asked me where I was, I would undoubtedly have replied Bloomsbury, not Norfolk. Hypnosis has the power to put you in two minds which, normally, do not so easily co-exist. Or perhaps it simply intensifies the imagined at the expense of the real. I knew I was imagining Norfolk, but I was doing so with heightened capacities of evocation and memory. The real – Bloomsbury – had simply become a blurred background.

In addition, I felt sure I could have resisted the imaginary balloon he had attached to my wrist, but I simply did not want to. Free will existed as surely as the sounds in the office, but it had been reduced to a distant background sound. Had I tried to tell myself to lower my arm, I simply couldn't have been bothered. This would have appeared to be a free choice, but, in fact, the indolence – the 'couldn't have been bothered' – was just my way of accounting for the perhaps unpalatable fact that my free will had been overruled.

Eventually, after the rising wrist and the glued eyes, Oakley relaxed me further and said: 'Now take the calm feeling to your special place, sitting on that bench in the sunshine.' I returned to Norfolk. Then he simply asked if I noticed anything odd about it. At once I did. About two hundred yards beyond the river rises a small hill topped with a dense thicket of trees. In the midst of these trees I now saw a glow. At first I tried to interpret this as a memory of a fire I had once seen up there. But that was at night and some way to the left. In any case, this glow was not fire-shaped, it was distinctly oval, flying saucer shaped in fact. I had made up my mind that it was, indeed, a UFO, but, just as I did so, Oakley quickly brought me out of the trance. He felt he didn't know me well enough to take me further and he didn't wish to add an overtone of anxiety to a place where I felt relaxed.

This didn't matter. What did matter was the clarity and certainty with which I saw that glowing UFO. It was, quite simply, there. Obviously, I knew of Oakley's past creation of abduction narratives under hypnosis and, equally obviously, UFOs were on my mind. I was predisposed to see this. But there is a big difference between thinking such things and seeing them. And I was definitely seeing this. There is, however, a nuance here. Remember, I knew I was in Bloomsbury and, therefore, I could not be seeing this now. I was merely evoking Norfolk in my mind. But is evoking the right word? Perhaps I was remembering it; maybe I was, indeed, remembering an occasion on which I had seen a UFO in Norfolk but my conscious mind had suppressed the memory or it had been suppressed for me. Or I was imagining it, imagining Norfolk with a UFO attached. I could do such a thing in a conscious state, but it would involve an effort that constantly reminded me it was not real. Under hypnosis, there was no doubt in my mind that I was really seeing this and I was curious to see more. Note felt I could *see* – i.e. be shown – more; I didn't feel I needed to *imagine* – i.e. make up – more. There was, in short, absolutely no feeling that I was generating the image myself.

Oakley had ensured that I came out of the trance feeling exceptionally well and relaxed – and I did – but I was also dis-orientated. I hadn't been prepared either to be deeply hypnotized or for the strange ambiguities of the experience. Of course, the simplest explanation is probably true. I had never actually seen a UFO, I had simply imagined one under three distinct influences: my knowledge of Oakley's experiments, my own preoccupations at the time and the effects of a deep hypnotic trance. But the point is: I didn't feel as though I was imagining anything. Listening to myself on the tape I made of the session, I can hear my voice sounds tentative, investigatory, not as though I am idly inventing something.

This means that, if Oakley had been determined to convince me that what I was seeing was a recollection and not an invention, then he could have done so, assuming I did not resist too fiercely. He would merely have confirmed what I felt to be true, that this was a flying saucer. Evidently, therefore, if a large number of people who are convinced alien abductions are real are hypnotizing even larger numbers of others who suspect they might be, then it is likely there will be many alien abduction narratives flying around, as, indeed, there are. Of course, this is not proof they are not true, but it does provide a persuasive context for a simple psychosocial explanation. Hypnotism is a technique that triggers a mass storytelling project in which all the stories are linked.

On the basis of my experience, I cannot argue with this. But I do believe it does not say enough. All it tells us – assuming we accept that the saucer was an artefact of the hypnosis – is that we are prone to see such things in such a state. It does not tell us why and it does not answer the deeper questions about the nature of human experience that are implied by the phenomenon of visions seen under hypnosis. These questions become even more profound when we remember that many people see these things in full consciousness and without any suppression of the memories they inspire. These are questions about the way we apprehend the world, not about the truth or otherwise of alien experiences. I shall come back to this at length.

But, for the moment, there is an important link to be made here. Many abduction stories surfaced at the same time as stories of satanic child abuse. This was the widespread belief that not only were children being abused, they were being ritually abused. Specifically, they were being drawn into devil worship, black masses and all the paraphernalia of a horror film. Fired, probably, by intense media interest, these stories became increasingly exotic. Women were being used as 'brood mares', breeders of children specifically to be abused. And there were tales of child sacrifice. Much of this was reinforced by therapists using hypnotic regression. Indeed, the use of this and other therapy techniques seemed to lead to the exposure of all forms of child abuse on a massive, almost all-pervasive scale. Many celebrities, it transpired, had been abused as children.

My own experience of satanic abuse, as a reporter, made it clear to me what was happening. This was when a case exploded in Rochdale, Lancashire, which had led to many children being taken from their homes by police. Within minutes of arriving in Rochdale, I knew there was no physical evidence and the testimony of the social workers – all the police had to go on – was useless, not to say illiterate. But I did discover that the social workers had been lectured by some Americans on satanic abuse and, within weeks of returning to Rochdale, they had found some. Here it was the social workers, not the children, who were victims of suggestion.

Defenders of hypnosis in these cases talked of 'recovered memories'. Sceptics swiftly responded with the term 'false memory syndrome'. The mania for satanic abuse claims died down – in itself, clear enough evidence that, in this case, the sceptics were largely right. It simply wasn't happening on any appreciable scale. But it also died down because it was a mythology with real-world implications. Accusing people of child abuse is a serious legal matter with potentially appalling consequences. Once it became clear that the social workers' suspicions were unsupported by physical evidence, the authorities ceased to give them credence. As a story, it died.

In the case of abduction stories, there are fewer real-world implications. People don't get arrested or have their children taken away from them. Alien cults such as Heaven's Gate are an exception in that membership can become life-threatening, but, on the whole, alien abduction is a relatively harmless affair, at least for those who are not abducted. As a result, fewer people are actively trying to suppress such revelations, and there is seldom a legal reason to do so.

Believers in abduction, like Budd Hopkins, say that the connection with abuse stories is wrong. Hopkins asserts that memories of abuse revealed during therapy are often false, and may even derive solely from the therapist. Alien abductees, usually suspect or know already that they have been abducted, and they just want to find out more about it. In other words, the abduction is not uncovered during therapy.

But, without judging that issue, there are generalized connections between the two phenomena. Both have high levels of media reinforcement. Satanic abuse had huge coverage and alien abduction has both been reported as news and been used as the basis for mass-market fictions like *Taken* or *The X-Files*. Both forms of narrative have, therefore, been very aggressively disseminated. Both involve the idea of hidden knowledge unsupported by adequate physical evidence and, in consequence, requiring a degree of faith. And both, therefore, create a coterie of believers mutually reinforced by a defensive posture to the disbelief of the outside world. Campaigners against Satanic child abuse and abductees are all convinced that the prevailing world view is fundamentally wrong, that something extraordinary and exotic is going on unseen by ordinary people in ordinary states of consciousness. Finally, both, of course, are heavily dependent on the idea of repressed memories that need to be retrieved by therapeutic means.

These are not, for my purposes, items of evidence for the truth or otherwise of either alien abduction or Satanic child abuse. They are statements about the human condition.

Hypnosis itself is an aspect of the mind control obsession of the modern world. This has manifested itself in the form of Cold War fears about 'brainwashing' and conspiracy theories about experiments with telepathy and behaviour control by the CIA and/or KGB. The feeling is that rapidly advancing technology cannot be stopped by the feeble barrier of our skulls. Somehow 'They' will find a way of getting inside. 'They' are often, of course, the aliens. In the film *Earth Versus the Flying Saucers* (Fred F. Sears, 1956), for example, an unfortunate major is subjected to the aliens' 'Infinitely Indexed Memory Bank' and is left wandering around like a zombie. The central idea is that what is in our heads is reducible to lines of computer code, an idea derived from the habitual depiction of computers as machine brains. The alien mind control in the case of abductions is the imposed amnesia. Human hypnotism is, in this context, a mind-freeing mechanism.

Under the heading of 'suggestion', there is one final story that should be told. This involves two sceptical attempts at a scientific explanation of why alien encounters and abductions appear to happen at all. I call these attempts 'sceptical' but that may be a little harsh. In neither case do the researchers take the extreme sceptical position that the experiences are not real, they are merely made up. Rather they accept the reality of the experiences, but do not claim they are caused by extraterrestrial intervention.

The choice of these two is partially arbitrary as there are many to choose from. One theory, for example, suggests it is caused by a memory of the trauma of being born. Or it is to do with the personality of the abductees, specifically their tendency to fantasize. Or it is to do with fantasies of masochism and escape from self. And so on.

Psychologists have become increasingly interested in the field of alien abduction for two reasons: there is a lot of it about and none of it, they believe, can possibly be true. The phenomenon is sufficiently widespread to be deemed significant and, as it cannot actually be happening, it must cast some light on dreams, fantasies, on our entire way of apprehending reality. That it cannot be happening is, of course, a statement of faith. Chris French of Goldsmiths College, London, put this faith most lucidly in his paper 'Alien Abduction Experiences: Some Clues from Neuropsychology and Neuropsychiatry'.

'If abductees are not deliberately deceiving other people, is it possible that they really were abducted by aliens? This seems highly unlikely, as the evidence presented in support of alien visitation is far weaker than uncritical and sensationalist media coverage typically implies. Even such celebrated cases as the alleged crash of a flying saucer and recovery of alien bodies near Roswell, New Mexico, in 1947 are in fact based on incredibly weak evidence.'

This may incense believers, but, in fact, the interest of main-stream psychology in aliens represents a significant improvement of their position. Once they were dismissed as deluded, mad, marginal or hoaxers. At least now their reported experiences, if not the causes, are being accepted as real. Indeed, in believing circles in Britain, French's position is accepted as honourable and sympathetic. He is, however, sometimes seen as a liability. He once visited the site of the Rendlesham encounter – 'Britain's Roswell' – where, nightly, parties still go out to watch for UFOs. When the woman who organizes these watches realized French was there, she simply said there would be no sightings that night.

'She said, "That's it, then, nothing will happen." I had this mental image of this fleet of interstellar spaceships calling to each other, "Chris French is there! Abandon mission! Abandon mission!" It was quite flattering.'

The two examples of scientific explanation I have chosen are not from the realm of pure psychology in that they involve the identification of distinct physical causes. These, it seems to me, demonstrate the kind of thought processes deployed by scientists in this area and, in an important way, the shape of the ideas involved.

The first is sleep paralysis. During REM (rapid eye movement) sleep, when we are dreaming, the muscles of the body are paralysed. There may be a good reason for this: if we weren't paralysed we might attempt to live out our dreams with dangerous consequences. Occasionally, people become aware of this paralysis and this condition is sometimes more accurately called awareness sleep paralysis (ASP). This is said to be rare, but, in fact, I'm not so sure it is. It has certainly happened to me – my sleep patterns have always been disturbed and chaotic – and, in Japan, it is an aspect of popular culture. There it is called *kanashibari*. In this the sleeper wakes, finds himself paralysed and is terrified to discover the presence of another being. There is plenty of popular discussion of this in Japan and, it is said, the experience offers an opportunity to communicate with spirits. It is significant that *kanashibari* is so directly associated with the idea of a presence in the sleeper's room.

In ASP the person is usually lying on their back and may be about to fall asleep or just waking up. These experiences half-way between sleeping and waking are called hypnogogic when falling asleep and hypnopompic when waking up. Even when ASP is not involved, they produce peculiar and frequently exotic effects. I find in the hypnogogic state that dream causality suddenly invades otherwise normal waking narratives. I might be thinking through the events of the day and suddenly fly from one place to another, or I might hear the sound of somebody in the room who simply cannot be there. If I wake up I become aware of the oddity, otherwise, I assume, it is lost in the descent into full unconsciousness.

ASP often involves the person feeling they are becoming heavier and heavier and there might be increased heart rate, breathing difficulties and feelings of anxiety or dread. This can last from a few seconds to a few minutes. Studies suggest that between 25 and 40 per cent of the population have been afflicted. Yet, as French points out, it is not known as a condition among the general public, at least outside Japan.

'People are thus unlikely,' he writes, 'to describe their unusual experiences in terms of ASP, but rather to describe it in terms of a spirit encounter, or an encounter with a being from another planet.'

This is probably true but it is also poignantly resonant. Without a scientific explanation, people resort to aliens and spirits. But, with a scientific explanation, that world is banished. The history of the modern world is captured in French's offhand sentence.

There are a number of elements of alien encounters that match the symptoms of ASP. They often happen at night, time passes in a strange way, the person is paralysed, they may happen repeatedly to the same person and they both involve seeing or being aware of strange beings. As ASP has long been undiagnosed and, in the public realm, undefined, then surveys like the Roper Poll might be entirely explained by its hitherto underestimated prevalence. The poll looked for peripheral symptoms that abduction had taken place. But these could also be symptoms of ASP.

Plainly, if ASP is indeed to blame, then this would explain both the prevalence of the experience and why abductees do not show other symptoms of mental disturbance. ASP just happens to people, it is not an aspect of a broader psychiatric condition.

But a good deal is still left to be accounted for, most obviously why do our brains have these peculiar eccentricities and why do they manifest themselves as alien presences?

Michael Persinger, a psychologist and neuroscientist, has suggested one answer. He suspects it may be something to do with patterns of activity in the temporal lobes, the areas at the front of the brain concerned with perception and memory. In particular, people with highly labile – unstable – temporal lobes may be more prone to abduction experiences.

Persinger's starting point for this argument is important. All human experience is generated within the brain. We may think we see through our eyes, but we don't. We see the effect of the stimulation of our eyes by light only once it has been translated into the language of the brain. This language is being spoken all the time, when we are awake and when we dream. Our awareness of the outside world is simply patterns forming within this chatter. It is not direct awareness of the things themselves.

This means that '*any* stimulus that can induce specific patterns of activity within groups of brain cells can generate experiences that are equally as real and compelling' as what we would call the real thing. If seeing a giraffe produces a certain pattern of response in the brain, then reproducing that pattern will make you see a giraffe even if no giraffe is present. Like French's sentence about the absence of an awareness of the scientific explanation of ASP, this seems to me to be an idea of some cultural poignancy and resonance.

Persinger is aware that, locked inside our brains as we are, some mutually recognized system must exist to keep us in contact with others and reasonably stable within society. Otherwise we would simply descend into trance-like self-involvement. That system is belief.

Beliefs are people's experiences of how they organize, predict, and explain the myriad random stimuli that impinge upon them. Without the structure-making properties of beliefs and a functional prefrontal region, normal people experience incessant anxiety, panic or feelings of impending doom. Without shared beliefs that there is a tomorrow, the complexities of cultural order and social conduct would deteriorate within days. When the majority of people within a culture share a belief about why they exist and what will happen to them in the future, the belief is considered normal. If a belief is not shared by the majority, it is labeled as delusional. The label does not reveal the source of the belief, nor does it help to elucidate its origin.

It has been known for centuries that people with electrically overactive brains report intense experiences and, writes Persinger, 'virtually every basic element of mystical, religious, and the UFO visitor experiences have been associated with spontaneous electrical seizures that are not necessarily associated with convulsions or obvious changes in behavior'. In fact, these experiences can be replicated by direct, artificial electrical stimulation of certain areas of the brain.

Abductions are similar to visitation experiences reported in the past – incubi, succubi, angels and fairies. All the elements of typical abduction accounts have been replicated by direct stimulation of the temporal lobes. One of the oddest themes in these accounts is the side on which the alien is seen. If seen on the left side, the alien will tend to evoke fear and apprehension. On the right side, it is more likely to speak or deliver a message.

Persinger has developed a system of magnetic fields with which he surrounds the head of an experimental subject in an attempt to provoke a visitation experience. In an article in *New Scientist* (19 November 1994), Susan Blackmore, a psychologist and physiologist, described the experience.

For the first ten minutes or so nothing seemed to happen. To tell the truth I felt rather daft. Instructed to describe aloud anything that happened I did not know what to say and felt under pressure to say something – anything. Then suddenly all my doubts were gone. 'I'm swaying. It's like being on a hammock.' Then it felt for all the world as though two hands had grabbed my shoulders and were bodily yanking me upright. I knew I was still lying in the reclining chair, but someone, or something, was pulling me up.

Something seemed to get hold of my leg and pull it, distort it, and drag it up the wall. I felt as though I had been stretched half way up to the ceiling.

Then came the emotions. Totally out of the blue, but intensely and vividly, I felt suddenly angry – not just mildly cross but that sort of determinedly clear-minded anger out of which you act – only there was nothing and no one to act on. After perhaps ten seconds it was gone but later was replaced by an equally sudden fit of fear. I was just suddenly terrified – of nothing in particular. Never in my life have I had such powerful sensations coupled with the total lack of anything to blame them on. I was almost looking around the little room to find who was doing it.

Plainly, even in a sceptic like Blackmore, these magnetic fields can induce intense experiences. It is Persinger's view that the persistence of the visitation experience through history indicates that there is 'a statistical sequence of activity during twilight states'.

'We suggest that the experiences of movement, a dream-like atmosphere, the inability to move, sensed stimulation of the anal or genital regions, feelings of "probing", and the intrusion of humanoid perceptions are experiences of a synthesis. The synthesis is due to the electrical coherence that occurs between regions of the brain that are partitioned or maintained functionally isolated during the normal waking state.'

The difference between aliens on the left and those on the right may, speculates Persinger, be due to a relaxation of the normal division between the right and left hemispheres of the brain. In normal waking states the left hemisphere is predominant, this is associated with the clear sense of self. But, in twilight states, the right hemisphere rises to prominence. This is associated with emotion, language, spatial detail and so on. In these states the self becomes more aware of the right-hemisphere activity 'not filtered or translated by left-hemisphere sequences'.

'We have suggested that during these transient intercalations, you become aware of the right hemispheric homologue of the left hemispheric sense of self, and you experience the presence. The presence is effectively your right hemispheric representation of the self.'

The ghostly presence in the room is your other self.

Sleep paralysis and temporal lobe lability are two important ways of finding a physical basis for the alien contact and abduction experience. They will console the sceptically minded, allowing them to say: these people might not be nuts but what they say is happening to them is not real, it is a product of their brains in certain disordered states.

Yet, as Persinger says, all that we know of the real is in our brains. We have no direct contact with the real. What we mean by real is what is in our brains. And yet we act on the basis of a real that is not in our brains, what we like to call the world. Steve Grand, the independent artificial intelligence researcher, has pointed out that this is, in fact, impossible. The physical mechanisms of input and output to the brain are simply not quick enough to enable us to do what we do. Using his example, shooting a bird in flight would be impossible if we were taking in information and then calculating each point of its trajectory. What we appear to do is make a virtual world in our brains that allows us to predict the bird's flight. This reinforces Persinger's point about all reality being in our brains and, in many ways, it overturns much conventional thought about artificial intelligence.

What is clear is that, in the realm of alien contact and abduction, it is the nature of the real that is at stake. And, if science's ongoing project is to disenchant the world, to subdue the strange, then in this area it has so far failed. The idea of a secret alien presence on Earth is, indeed, strange. But is it so much stranger than the idea that our consciousness is naturally prone to these visions, that the visitor in the bedroom is a fact of human life, that the world we know is not, in fact, the world in which we live? And finally, if it is not, what is it?

Answers must now leave the stage to be replaced by questions.

PART TWO

QUESTIONS

11

Claudius Ptolemaeus/
Galileo Galilei

The eyes of Claudius Ptolemaeus, better known simply as
Ptolemy, opened circa AD 90, possibly in Hermiou in upper
Egypt, and they closed finally some eighty years later in the city
of Alexandria. Ptolemy is a Greek-Egyptian name and Claudius
Roman. Possibly his family was Greek and he was a citizen of
Rome. In birth, he seems to have embodied the classical roots of
Western civilization. In life, he was to become one of its strongest
and most enduring branches.

Ptolemy's eyes were among the most effective that ever looked
at the night sky. He was a magus of sight or, as we would say, an
astronomer. His first book on the subject is generally known by
its Arabic-derived name of the Almagest. Its original title in
Greek meant the Mathematical Compilation, this was changed to
the Greatest Compilation. In Arabic this became *al-Majisti*,
which, in turn, became Almagest when it was translated from
Arabic to Latin. The book, like its maker, was protean, pan-
cultural.

The Almagest is a work in thirteen volumes that told people
how to use their eyes when they looked upward at night. It did so

with such overpowering authority and precision that its interpretation of the movement of the heavenly bodies was not to be
seriously challenged for 1400 years. This should not be seen as a
book about astronomy. For most of the history of the West, it
was astronomy.

The reason it was so eagerly accepted as a universal
orthodoxy was that it provided a scientific and observational
basis for the unification of the classical and Christian understanding of the cosmos. Unifying classical and Christian wisdom
in all areas, not just astronomy, is a puzzle that has always
dogged theology. Solving it is rather like the attempt to unify
relativity and quantum theory in our own day. It should be
possible but it isn't. They both appear to be true but they
contradict each other. Classical thinkers seemed to have thought
the greatest thoughts and classical artists seemed to have made
the greatest things. But they predated Christ. They were
unbaptized. The further embarrassment was that, for a long time
after Christ, Christendom failed to compete. The Roman Empire
turned Christian long after its greatness had passed and the
ensuing Dark Ages were, well, dark. Islam, in the interim,
seemed to have seized the torch of civilization and scholarship.
This shouldn't have been happening.

So classicism was true by virtue of its achievements and
Christianity was true by virtue of revelation. Purgatory was one
compromise solution. In *The Divine Comedy*, the poet Dante is
escorted around the underworld by Virgil, the great poet of
classical antiquity, but it is made clear that Virgil must remain in
Purgatory. He is one of the unbaptized.

This solved the salvation issue – the classical masters were
great but not actually in heaven – but it did not solve the issue of
truth. The big problem was Aristotle. As a philosopher and
scientist, he had no peers. He was beyond question the most
brilliant man that had ever lived. His physics went unchallenged,
as did his cosmology. At the heart of his understanding of the
heavens was his view that the Earth was at the centre and all the
other heavenly bodies orbited around us fixed in a series of
concentric shells. The matter of the Earth was impure, mutable,
but, from the moon outwards, all was unchanging purity. This

is understandable. That is what, from the unassisted human perspective, appears to be the case.

Some sort of synthesis was required to fit all this together. This was achieved by the Christian humanists of the Middle Ages. Essentially, they said that Aristotle was right and that he had glimpsed the divine truth. His cosmology fitted with the Christian view that humans were the point of the whole exercise. The stars and planets circling above us were about us, they were an aspect of our journey to God's kingdom.

Unfortunately, their bizarre movements did not always seem to confirm this geocentric hypothesis. But Ptolemy had, long before, found a way to make them do so. The Almagest is, in essence, just a fantastically ingenious mathematical solution to the problem of an Earth-centred universe. Once Ptolemy's had been accepted as the correct view of the cosmos, the correct way to see at night, by the medieval scholars of the church, refuting him became heretical. As a result, in the millennium and a half before Ptolemy was finally refuted, every new observation of the heavens was fitted into his scheme. It became a vast, bloated, complex of proliferating epicycles, little orbits inserted to shore up the big picture.

There is an important point about this big picture. In spite of its complexity, it is a very literal view. In effect, the Ptolemaic orthodoxy had taken what we see in the hemisphere of the heavens at face value. It told us to believe our eyes and our eyes told us that we were at the centre of this system of circulating bodies. Everything revolved around us. We were the point of the whole exercise. The sun really did rise in the morning and set at night. The language we used about the world was not just metaphorically apt, it was true. Of course, the stars and planets may pursue more complex paths than the sun and the moon; but, still, we were at the centre. We habitually see what Ptolemy saw and, for centuries, we believed what Aristotle and Aquinas told us it meant. Seeing was indeed believing.

But it was all a trick of the light, a bug in our programming, an acutely painful historical reminder of the partiality of our perceptions. Any observer will always be at the centre of his field of vision. It is a mere geometric fact and certainly no way to build a universe. But that's what we did, we constructed a cosmos out of our eyes, we turned seeing into believing.

Of course, finally, as more observations were made, the Ptolemaic system began to buckle under its own weight. To keep the Earth at the centre required cosmic movements of insane complexity. By the sixteenth century the supposed rationality of the epicyclic dance of the planets and stars had become all but indistinguishable from chaos. Surely there was a simpler explanation. What if, asked Copernicus in 1543, the sun was at the centre? Wouldn't everything be so much easier? But he only asked. Galileo answered. In 1609, he applied a prosthesis to his eye, a telescope, becoming a cyborg – part machine, part organism. Indeed, he became the defining cyborg of modernity. Cyborgs have superhuman powers and Galileo, superhumanly, saw what he was not allowed to see, what no human had ever seen – mountains on the moon, moons around Jupiter. They weren't supposed to be there. The Ptolemaic system and the whole of classical and Christian cosmology crumbled. Ptolemy, Copernicus and Galileo had all *looked* at the same thing – the heavens – but they had each *seen* something different. We now have Galilean eyes.

Galileo, having been at first persecuted by the church and then, in death, forgiven, prevailed. He became the founder of modern science. By subjecting all authority and theory to the test of experience, he had created the experimental method. Ptolemaic astronomy was based on the assumption that if any observation didn't fit the theory, then the observation was wrong. Galilean astronomy assumed the opposite. If Jupiter had moons and the theory said it couldn't have, then the theory was wrong. Glaringly obvious as this may seem to us today, it was a revolution in human thought whose effects we still cannot fully comprehend. It changed everything. Contemporary politics, for example, are Galilean. People do not believe what they are told by authority: they believe in their own private experiments with reality. This leads, of course, to the distinctive issue of post-modernity – that reality does not accord with the results of our experiments. At this point the Galilean revolution seems to exhaust itself.

But the simple, basic effect was the revelation that it only *looked* as though we were at the centre of the cosmos. In reality, we were not, far from it. If we were in the middle of anything, it was nowhere. Lost, in fact, in space. And then, when Newton found the celestial clockwork by which this space operated, things became even worse. Not only were we lost in space, we were an insignificant cog in a machine which could happily run on without us.

This new sense of mere optical centrality is cold and lonely. The old Aristotelian-Ptolemaic-Thomist universe was flattering, it was arranged about us to endorse our ultimate value, our primacy in the cosmos. Next to that, the primacy offered by the scientific age is a trivial thing. We just happen to be chance agglomerations of matter, organized so as to sustain the ability to see and think. Of course, we cling to our old ways of emotion, feeling and aspiration. We find the moon romantic, the stars sublime. The sun still 'rises' in morning hope and sets in evening serenity. We still look out at the cosmos with soft, human eyes. We remain victims of what literary critics like to call the Pathetic Fallacy – the delusion that nature is in accord with our moods and inclinations. But then Galileo cyborged himself and a new, alien gaze began to regard the world, and saw at once that this was not home.

Experiment on a Bird in the Air Pump in the National Gallery, London, was painted by Joseph Wright of Derby in 1768. 'A travelling scientist,' explains the gallery's note, 'is shown demonstrating the formation of a vacuum by withdrawing air from a flask containing a white cockatoo, though common birds like sparrows would normally have been used.' One small girl looks on in horror at the dying bird, another averts her gaze. A teacher tries to explain the significance of the experiment; a philosopher ponders in the corner. A pair of young lovers have eyes only for each other. Through a window the moon is visible.

Staring beyond all this are the eyes of the Galilean experimenter. He is not looking at the bird, he is not looking at the other people in the picture and nor is he looking at us. The troubled girls expect an answering, sympathetic glance to allay their fear of this bird murderer, but he does not oblige. He is seized by a new knowledge of reality, an understanding beyond lovers, birds or girls. His eyes are alien, he is possessed, ordinary human emotions mean nothing to him. He is what we would come to know as the mad scientist.

Baron Frankenstein in Mary Shelley's novel of 1818 is, in the same terms, mad. He is the romantic scientist who creates an artificial – therefore alien – creature out of spare body parts. The very act says that the human body and, by implication, the human self is a reasonable thing, a mechanical construct. It is a romantic act but it goes wrong. Shelley's artistic romanticism exacts revenge on the scientific version. This vengeance is an aesthetic act too perfectly attuned to the modern imagination ever to be forgotten. Frankenstein, the creator of aliens, constantly reappears as the mythic rebuke to scientific hubris. In Greg Bear's 1985 novel *Blood Music*, a scientist wearily considers the limiting, moral effects of this legacy:

> And Michael Bernard knew all too well the frustrations of being stopped dead in his tracks while following a promising path of research. He could have cured thousands of people of Parkinson's disease . . . if he had simply been allowed to collect brain tissue from aborted embryos. Instead, in their moral fervor, the people with and without faces who had contrived to stop him had also contrived to let thousands of people suffer and be degraded. How often had he wished that young Mary Shelley had never written her book, or at least had never chosen a *German* name for her scientist. All the concatenations of the early nineteenth and mid-twentieth century, coming together in people's minds.

The rational, romantic impulse to transform the world through the application of scientific reason wearily confronts the irrational impulse to transform the world through the artistic imagination. The former embraces the alien, the latter rejects it. Professor Bernard Quatermass, the fabulously cantankerous and probably mad British government rocket scientist on television and in films in the fifties and sixties, always regards human feelings as trivial when set against the great scientific project to know, to see, the universe. He welcomes aliens – extraterrestrial as opposed to artificial – even when they threaten to destroy human life. And he does so because the project must go on. It is why we are here. And Dr Arthur Carrington, the scientist in *The Thing from Another World* (Christian Nyby, 1951) demands that a murderous alien be protected because knowledge is more important than life. Knowledge, after all, endures and accumulates, lives do not.

What such mad scientists see is what is real and what is real, has no time for or interest in the tears of little girls. Such things barely exist at all in the great scheme of things. They are no longer at the centre of Ptolemy's universe. The idea that they should get in the way frustrates Bear's Michael Bernard and infuriates Quatermass. The space encompassed by the gaze of the impassioned scientist is so vast that, to him, we are flecks of dust, persisting for a mere moment and changing nothing because nothing can be changed in the great cosmic machine, only observed. And the laws he detects working behind this vastness do not explain pity. Why should they?

The horror of the mad scientist's gaze derives from our dawning awareness that the mind behind the eyes has been evacuated of all human feeling. We cannot appeal for respite or mercy, for the eyes signal there is none to be had. These eyes are alien to the empathetic and the emotional. They are, in this sense, non-human. They are eyes that have seen the vanity and delusions of humanity for what they really are. Nothing. The one human impulse worth preserving is the rage to know, but not in order to transcend our condition, only so that we will better understand the exact nature of our emptiness. Note that Wright paints a 'travelling scientist', this man has no need of a home. Like so many other heroes of modernity, his heroism is founded on perpetual mobility.

The girls in Wright's painting cling on to the picture the experimenter denies them. They cannot do otherwise. Human eyes are forever gazing at two quite different worlds – the one that concerns us of Ptolemy and the one that does not of Galileo. It is the difference between day and night. Our days are full of ourselves and our projects, our nights teem with otherness.

In our daily, practical, useful lives, we see the world of the approximate disc defined by our field of vision. On the disc things make sense. Everything lies at an understandable distance. We can reach these things. For a long time we could not reach the clouds, but sometimes they came down to the ground and, anyway, they always seemed, like us, to be afflicted by change and chance, earthly forces. This disc is our home.

But the disc is only the base of the snowstorm paperweight, the bell jar, in which we live. For above us is the hemisphere of the heavens. This is an affront to our domesticity. By day it is a fathomless blue, by night an unconsoling black, dotted with heavenly bodies that move on, regardless of our gaze, indifferent to our prayers. They are indifferent, also, to our desire to explore. The stars are unreachable, shining from some incalculable distance.

We know the disc and just above it pretty well. We can reach out and touch these things. But the hemisphere is beyond us. We cannot touch it and we can only speculate about what it contains. We see its lights and, with the development of radio astronomy, we hear its voices. But they do not speak to us; they are not with us; they preceded us; they will outlast us.

Worst of all, we now know the extent of that awful darkness. The figures dazzle and daze. Proxima Centauri – the 'closest' star after our sun – is 4.2 light years away, 1.29 parsecs or about 40,000,000,000,000 kilometres. There may be 100,000,000,000,000,000,000 other stars in the universe. We can look at, but we cannot see, such distances and such numbers. Even that suavely paradoxical phrase 'light year' insults our visual sense of scale. These quantities are brutal, oppressive, more than dazzling, they are *blinding*.

Worse still. We look out on the disc that is our home and millions of eyes return our gaze. We look up at the hemisphere and nobody and nothing looks back. In their immensity, the heavens are not only blinding, they are blind. This is the message we have taken from modern science, that we are small, alone and little more than a chemical accident.

'If he accepts this message in its full significance,' wrote the biologist Jacques Monod, 'man must at last wake out of his millenary dream and discover his total solitude, his fundamental isolation. He must realize that, like a gypsy, he lives on the boundary of an alien world; a world that is deaf to his music, and as indifferent to his hopes as it is to his suffering or his crimes.'

Monod is using the words 'deaf' and 'indifferent' as metaphors. He might have added 'blind'. In reality, his universe is none of these things. It is just nothing. And what he means by 'alien' is that nothingness. What could be more alien to self-conscious man, the supreme something, than nothing? Monod wants us to wake up to the fact that nothing will ever look back from that terrible void and tell us we matter.

But how can we accept this message, agitated as we are by the hemisphere above? The sight of the heavens makes us what we are – creatures troubled by existence in its every aspect. The fact of the hemisphere of the heavens is the source of the enervating extremity that haunts human existence. We think we are upset because of love or war, death or art, creed or colour, or because of the seeming intensity yet clear futility of our lives. But we aren't. We are upset because of the things that float above us and because of that indifferent blue or damnable black. We are incapable of being lukewarm on the subject of the sun, the moon, the stars, comets or meteors, pulsars or quasars, that infinite blueness, that utter blackness. The heavens demand an agonized or ecstatic response. They give us no peace. Somehow, we cannot imagine they are as blandly indifferent to our existence as they seem. Somehow, we feel, there must be eyes out there somewhere.

12

Alien Eyes

Imagine that we can see nothing of the heavens. Perhaps Earth is shrouded in permanent cloud. Or, perhaps, some optical trick in the upper atmosphere makes the sky as opaque at night as it is during the day. Assume this cloud or opacity is too dense for aircraft to penetrate and its very existence means that the construction of spaceships has never occurred to us. Where would they go? We never see the sun, the stars, the planets, the moon, the darkness of space. We know nothing of their existence. There is no beyond. Ask yourself: what difference does it make? Probably it means we can sleep better, never having to fear the spectacle of the appalling emptiness that hovers over us every night. But what else? We are, perhaps, happier, more at peace.

On Wesker's World, in Harry Harrison's short story 'An Alien Agony' (1962), the Weskers, small amphibious creatures with ribbed ears that fold like bats' wings, live a life of dull but relentless logic. They are intelligent, but they seem to be crawling up the developmental track with little passion or involvement. Perpetual cloud covers their world, the Weskers can see nothing beyond. They do not ask much of their world, only that it continue in much the same way in the future as it has in the past.

In this, they are to be disappointed. One day, a human, Trader

Garth, bursts through this cloud cover in his 'sky ship'. It is, for the Weskers, a Copernican moment. Their planet is not all that there is. There is something beyond. The Weskers realize they must adjust to a new reality. But they will do so slowly, as is their nature.

Garth, an aggressively secular man, likes the Weskers. Perhaps alone among the intelligent inhabitants of the universe, they have no god or gods, no beliefs, nor even superstitions. They are, in this, somewhat like the equally nice but equally dull Lithians in James Blish's *A Case of Conscience*. Their world is, to the Weskers, no more than what it appears to be – the place where they are born, feed, procreate and die. What more needs to be said? Thanks to the unbroken cloud, there are no stars to be seen, there is no cosmos to inspire speculation and stir ambition. The Weskers do not wonder what their existence is 'all' about, for 'all', if it is to mean anything, must imply more than 'this'. And, beneath the cloud, there is only 'this'.

Similarly, in Douglas Adams' novel *Life, the Universe and Everything*, the planet Krikkit is surrounded by a huge dust cloud. As a result, its inhabitants act 'as if they had a blind spot which extended 180 degrees from horizon to horizon'. They do not look up, they live lives of peaceful, harmonious con-templation. They seem more spiritually active than the Weskers or the Lithians, but they are equally complacent.

Then, once again, a spaceship plunges through the dust cloud and crashes on to the surface of their planet. In shock, the people of Krikkit immediately build their own spaceship – they are technologically more adept than the Weskers – to take them beyond the cloud to see 'the starry sweep of the Galaxy . . . the infinite sweep of the Universe'.

'It'll have to go,' conclude the men of Krikkit at the sight of this appalling, star-strewn emptiness. They immediately launch a devastating intergalactic war with the intention of eliminating every other life form in the cosmos. They see it as the only way to restore their peaceful, harmonious existence.

Krikkit and Wesker are utopias of visual ignorance. They have no need for a Ptolemy or a Galileo for there is nothing to explain or justify. Prior to the arrival of the spaceships, peace reigns on both planets because the heavens are invisible. The planets are hippie paradises, untroubled by excesses of ambition or the anguish of constant striving. And why are they so peaceful? Because they do not see the awful spectacle of the heavens. These beings are not agitated by the spectacle of an unknown that constantly threatens also to be unknowable. As a result, they live at peace with each other and in harmony with their surroundings.

So the Harrison–Adams solution to the thought experiment is that we would be much nicer, happier creatures if we lived beneath a canopy of perpetual cloud. The implication is that seeing the heavens troubles and agitates us, makes us bellicose and anguished. More specifically, since the creatures on Krikkit and Wesker are remarkable for their complete absence of ambition, it must be the sight of the stars and the black distances of space that condemn us to a constant struggle to improve ourselves or the world, to find new places to go. Looking up is the original sin. Like the apple in Eden, it exposes us to the agony of who we really are, the agony of the alien perspective in which we become alien to ourselves. We are never at peace and we wage war on nature because, for us, the universe never goes away. It is an unscratchable itch, the great anomaly, the monstrous absurdity with which we must live. Daily it reminds us that there is always much 'more' than 'this'. Seeing is believing. But how are we supposed to believe in this?

In a sense, what happened on Krikkit and Wesker happened to us in 1609. We did not have cloud cover prior to that, but we did regard the heavens as limited in extent and supportive of our self-esteem. We had created our own cloud cover. We need not feel vertigo when looking up.

But then, suddenly, we did. The Copernican and Galilean revolution did not simply displace the Earth from the centre, it also introduced the idea of infinite space. Without the Aristotelian shells, there was no reason for it to end. We were lost not just in space, but in infinite space.

Not surprisingly, confronted with this immensity, we began to have visions. But, if looking is the first sin, then the second is seeing things that aren't there. To the scientific mind, prior to Galileo everybody saw something that wasn't there – a geocentric cosmos. Of course, they also saw angels and demons, but the primary delusion, to the mind of modern science, was the Earth-centred universe. Today, this same scientific mind argues that that people who see UFOs are really seeing something else – most commonly, a weather balloon, Venus, a Stealth Bomber. They say that people who meet aliens are dreaming, fantasizing or victims of post-hypnotic suggestion. Only occasionally do they say that these people don't see anything at all, that they are hoaxers, that they've consciously made it all up. Generally, they accept that something has been seen. But they argue that it is just the seers' belief system that converts these things into inter-galactic starships or black-eyed Greys, just as the pre-Copernican belief system converted the universe into a gigantic dance around Earth. Of course, ufologists and abductees reply: we have seen them, therefore they are true. They are not really believers at all because they know; they have seen these things with their own eyes and seeing is believing.

But what we see, what we think we see and what we know are all facets of just one aspect of the alien encounter. The other aspect is the experience of *being seen* by intelligent, non-human eyes. Human vision has been displaced from its position of privilege. We know we are not at the centre any more; we suffer, as H. G. Wells put it, from 'a sense of dethronement'. But being dethroned in an empty, eyeless universe is one thing. Being dethroned in a universe that looks back is something else entirely. Being seen again raises questions of knowledge and belief. But, if we are being seen by alien eyes, then there is the further disorientating issue of how we are seen. What are we to them? What do they see?

Whitley Strieber wrote of an encounter with an alien female:

> Her gaze seemed capable of entering me deeply, and it was when I had looked directly into her eyes that I felt my first taste of profound unease. It was as if every vulnerable detail of myself were known to this being. Nobody in the world could know another human soul so well, nor could one man look into the eyes of another so deeply, and to such exact effect. I could actually feel the presence of that other person

with me – which was disturbing as it was curiously sensual
. . . To an intelligence of sufficiently greater power, it may
be that we would seem as obvious as animals seem to us –
and we might feel as exposed as do some dogs when their
masters stare into their eyes.

The shift in perspective – from seeing to being seen – is almost
intolerable. We become as animals under the alien gaze. This is
the great, the decisive passage from *The War of the Worlds* in
which H. G. Wells evokes this appalling sensation:

> For that moment I touched an emotion beyond the common
> range of men, yet one that the poor brutes we dominate
> know only too well. I felt as a rabbit might feel returning to
> his burrow and suddenly confronted by the work of a dozen
> busy navvies digging the foundations of a house. I felt the
> first inkling of a thing that presently grew quite clear in my
> mind, that oppressed me for many days, a sense of
> dethronement, a persuasion that I was no longer a master,
> but an animal among the animals under the Martian heel.
> With us it would be as with them, to lurk and watch, to run
> and hide; the fear and empire of man had passed away.

Perhaps this is literally correct. 'We're property,' said Charles
Fort, an eccentric connoisseur of the absurd, we belong to others.
One explanation for the apparent, to mainstream science at least,
silence of the heavens is the zoo hypothesis. This is that we are
being kept in what is, in fact, a perfect zoo, one in which there is
no interaction between keepers and animals. Benevolent beings
are watching over us but they have ensured we cannot see them.
We may be too dirty or dangerous or there may be an intergalactic rule that we cannot be allowed to see out of our zoo until
we have overcome our messy and belligerent natures.
  The further development of the zoo hypothesis is the peer
hypothesis. This suggests that there are superbeings who can
create their own universes, much as one would create a zoo. If we
actually lived in such an artificially engineered universe, the gulf
between us and its makers would be unbridgeable. If one such

superbeing created our universe, all intelligences in this universe would be similar to each other and to their creator – hence the use of the word 'peer'. The universe would be full of civilizations at more or less the same stage of development. As a result, they would not yet have achieved the ability to communicate with each other.

'Given mankind's present ability to communicate to the stars,' writes David Lamb, 'and the likelihood of interstellar travel by 2200 AD, then assuming Earth is typical of other habitable regions in the universe with a similarly developing timescale, we should expect to meet our peers fairly soon.'

But, of course, they also would be in the zoo, being watched. Could we bear to find out? The Harvard biologist George Wald thought not. At a conference in Boston in 1972 – Life Beyond Earth and the Mind of Man – Wald agreed that the universe could be full of life but added that he could 'conceive of no nightmare as terrifying as establishing such communication with a so-called superior (or, if you wish, advanced) technology in outer space'.

It is one thing to create our own technological wonder, but to be told how to do so by others would be traumatic.

'Just to get such information passively from outer space through that transmission is altogether different. One could fold the whole human enterprise – the arts, literature, science, the dignity, the worth, the meaning of man – and we would just be attached as by an umbilical cord to that "thing out there."'

Wald may be right. Nevertheless, alien lore is dominated by this determined search for alien eyes that return our gaze. Both in seeing aliens and being seen by them, there is the feeling that something fundamental, something to do with the roots of our identity, is at stake. This is because our eyes are more than just sensory organs and sight is more than just a sense. Touch, taste, smell and hearing are all, more or less, specialized. Taste is important when we eat and hearing when we listen to music. But sight is important at all times and on all occasions. The eyes are the windows of the soul, we say, implying that not only do they see, they can also be seen *through*, they can be penetrated. Or they can penetrate. The way somebody looks at us always means more than the mere fact that they are seeing us; it means that they have seen into us in some way or that they are signalling to us.

The eyes of aliens are different. They can penetrate in novel, disturbing ways. But they cannot be penetrated precisely because they are alien. We cannot know what they are thinking because we do not know how they think. We can only know that they are looking. Typically, the alien eyes are large and black, lacking the colour and detail of human eyes. Frequently they reflect what they see, like mirrors. They haunt and baffle those who see them. Seeing is believing; but what do *they* see and believe?

The idea of an utterly alien visual perspective is essential to an understanding of the alien encounter. The eyes of the aliens are almost always described in chilling or comforting detail by abductees and they are always mentioned by SF films and books. Eyes, for example, are a sign of assimilation in *Dune* (David Lynch, 1984), the film of Frank Herbert's novel. The eyes of the Fremen on the planet Arrakis are entirely blue because of the absorption of the spice melange which occurs only on this planet. Paul Atreides is assimilated by the Fremen and becomes Usul Muad'Dib, the Messiah. His eyes take on the blue glow of the Fremen, an alien visionary gleam.

And abductees frequently feel themselves possessed by the alien gaze. David Jacobs explains the process:

By using the optic nerve, the alien can, in effect, travel down the brain stem, into the autonomic nervous system in the spine, and then branch into the parasympathetic nervous system, giving him contact with virtually any organ. Abductees often talk about feeling physical sensations in their genitals, bladder, or other areas when an alien performs Mindscan procedures. The physiological responses necessary for erection and ejaculation in men, and tumescence, expansion, and lubrication in women can be artificially generated in this manner.

Or, as John Mack puts it: 'Finally, it must be noted that it is through contact with the aliens' huge black eyes, sometimes spoken of as like looking into a vast, all-knowing or engulfing void, that the most powerful feeling of connection occurs. The beings' eyes seem to provide a kind of intimation of the infinite, appearing to contain or evoke the sense of ancient unfathomable depths.'

In Ian Watson's novel *Miracle Visitors*, an abducted boy remembers the alien eyes. 'Their eyes are like Chinese eyes with hardly any eyelids. They're much longer than normal eyes. They go right round the corners of their cheeks, as though their eye sockets are a different shape from ours.'

In Pat Cadigan's cyberpunk novel *Mindplayers*, it is the eyes that have to be removed before the players can plug into the virtual reality machine. Once the eyes are engaged, the whole person is engaged. And in Greg Bear's *Blood Music*, the creation of billions of intelligent human cells alters physical reality. Picking up on one of the strangenesses of quantum theory – that observation changes the thing observed – Bear has his 'noocytes' controlling the world just by looking at it. A Soviet nuclear strike against the US is averted because the noocytes ensure that none of the missiles go off. New eyes engender new realities.

In the real world, after the wave of UFO sightings in 1947, the idea gripped the human imagination that, at last, we were about to experience a returning gaze. We must look up. But we had no idea what we were looking for. Flying saucers probably, but what did their pilots want? Perhaps they wanted, like Wells' Martians, to destroy us.

'Every one of you listening to my voice, tell the world, tell this to everybody wherever they are. Watch the skies, everywhere! Keep looking. Keep watching the skies . . . '

It is 1951 in the Arctic and reporter Ned 'Scotty' Scott is filing his story. A flying saucer has crashed into the ice to the delight of a US survey team.

'It's round!'

'We finally got one!'

'We found a flying saucer!'

But the saucer's occupant turns out to be a monstrous vegetable man. A battle ensues. The monster is finally killed by electrocution. There must be more where he came from and so Scotty must warn the world – 'Watch the skies, everywhere!'

In *The Thing from Another World* (Christian Nyby, 1951),
the skies deliver this vegetable man to Earth. Perhaps, then, it is
as well that the universe is blind – better no eyes at all than those
of this lumbering, crazed destroyer. But we must be vigilant and
heed Scotty's warning to watch the skies. In *The Day the Earth
Stood Still* (Robert Wise), the other big alien arrival film of 1951,
the alien was angry with us because of our sins. In Nyby's movie,
he's just angry.

But, then again, the thought of no eyes at all is terrifying if we
attempt to imagine space as nothing more than an unseen and
largely unseeable emptiness. Can it be possible that ours are the
only intelligent eyes gazing outwards at this immensity? Can it be
true that nothing has ever, will ever, look back and 'answer' our
gaze? Perhaps even the vegetable man is more comforting than
this.

We will never stop watching the skies, looking and wondering,
hoping and fearing. We never could; we didn't need Scotty to tell
us. Every human age has viewed the heavens as a puzzle picture
that must be solved. It was solved by Ptolemy and then by
Galileo. What we now see, in vastly greater detail, is what Galileo
saw: a universe in which we are a very small and seemingly
incidental part. We are not at the centre of anything. We revolve
around a pretty average star which is located at a not very
interesting point on a spiral arm of a perfectly routine galaxy.
There are trillions of stars, probably trillions of planets and
definitely billions of galaxies. All that seems odd about our planet
is that, for a desperately brief period, it has been covered with a
thin organic film called 'life'. And, of course, even this may not
be remotely odd or special. The universe may be full of life, but
the distances involved may well mean that we shall never know.
We shall simply have to carry on with the working supposition
that we are alone. If the universe has a plan for humanity, it
appears to be to convince us, beyond any possibility of doubt,
argument or, indeed, faith, that we really are nothing special.

But we have not quite shaken off the feeling that we are, somehow, special. We still see something of what Aristotle and Saint
Thomas Aquinas saw: ourselves at the centre of the universe, not,
this time, because God put us there but because we seem to be the

only ones doing the looking. We are still at the centre of our field of vision and there is no other field. We know of no other creatures that watch the universe and speculate about its nature, only human eyes and minds do that. We gaze outward but no one gazes back.

At the end of the mad scientist tradition, the one that began with Joseph Wright's staring experimenter, comes Hal, the computer in *2001: A Space Odyssey* (Stanley Kubrick, 1968). Hal has a single red eye, a glowing light behind a convex glass which reflects the world he sees. He controls all the systems on a spaceship, the *Discovery*, bound for Jupiter. This includes looking after the crew members who will be kept in deep sleep for the duration of the voyage. Hal is a single, intelligent entity but he is dispersed about the ship in the form of large black boxes, each with its own red eye.

The eye is utterly expressionless. It is, we assume, how Hal 'sees', though, until a crucial moment in the film, that is not made clear. That crucial moment is when two crew members lock themselves in a pod so that Hal cannot hear what they are saying. The camera then switches to a shot of what Hal sees. He is watching their mouths. He can lip read. There is no escape from his gaze.

Hal is the mad scientist who really does go mad. There is – though this is not fully explained until the sequel *2010* (Peter Hyams, 1984) – a conflict in his programming. On the one hand he is required to protect the crew, on the other he must carry out the ship's mission which is not, in fact, known to the crew. The mission overrides his duty of care and, in the madness created by this tension, he feels obliged to kill all of the crew except for one who survives to turn him off. Hal is the machine that does not stop, but he moves in pursuit of two purposes that pull him in opposite directions.

The eye of Hal does not change throughout, but we feel as though it does. At the beginning, his bland voice and personality suggest his gaze is steady and reassuring, unfeeling in the sense of reliable. But, as he cracks, the unfeeling gaze turns threatening. Like Joseph Wright's travelling scientist, he has become a being who cannot be 'reasoned' with in human terms because he is answering to a higher, or at least different, logic. The red eye of Hal the reliable guardian is only mildly alien. It is the eye of the intelligent other, but this other is one of us. He is our creation and our servant. But the red eye of Hal the psychotic is fully alien in that, when we try to see what lies beyond its gaze, we see a mind working on utterly different principles. The irony is, of course, that the mission which drives Hal mad is the pursuit of an alien – extraterrestrial – contact.

There are two aspects to the eye of Hal: first, it sees through human beings well enough to be able to thwart their plans and, secondly, the human beings cannot see through it. Hal's behaviour is indecipherable to the crew. The alien eye penetrates but cannot itself be penetrated.

From Galileo, via Wright's travelling scientist, Frankenstein and Quatermass, to Hal is a straight line, but not a necessary one. The gaze grows colder, ever more objective. But not always. We may now say that Ptolemy was deluded, that we have better instruments, more accurate observations. When we look though our telescopes, we see the real, not some self-serving fantasy.

It is not that simple. Telescopes do not save us from the pre-Copernican sin of seeing what is not there.

When Giovanni Schiaparelli died in 1910, he was celebrated as 'Italy's greatest scientist' and 'not only the greatest astronomer of Italy, but one of the greatest astronomers of our times'. A conscientious, cautious man of immense range, Schiaparelli was an emblem of the sober, objective pursuit of scientific truth. But he made some odd mistakes. His calculation of the rotation period of the planet Mercury, for example, was wildly wrong. But it was wrong for a reason. He wanted the planet to have seas, moderate seasons, a benign atmosphere; he wanted, in short, Mercury to be alive. His rotation period – eighty-eight days, the same as its orbital period – would have made this more likely than the true period, which is fifty-nine days.

Schiaparelli's objectivity was compromised by his cosmic pluralism – his firm conviction that we shared our universe with many other life forms, that even our closest planetary neighbours were inhabited. And so, when he observed another planet, Mars, yet again he saw evidence of life. In 1877 he told the world that he saw *canali* on the Martian surface. The word is usually employed to mean channels but could also mean canals. 'Channels' could indicate natural formations; whereas 'canals' suggests intelligent activity. Either way, Schiaparelli was convinced the straight, dark lines he saw indicated the work of a Martian civilization. And, as *canali* was always translated as 'canals', the English-speaking world came to share his conviction.

And nobody shared it more fervently than a Boston businessman named Percival Lowell, the man at the centre of what the historian Steven J. Dick has called 'surely one of the strangest and most contentious episodes in the history of science'. A brilliant, aristocratic figure, at the age of eleven Lowell composed 100 lines of Latin hexameters on the loss of a toy boat. At fifteen he set up his first telescope, a 6-cm refractor, on the roof of the family house in Brookline, Massachusetts. At Harvard he became known as 'the most brilliant man in Boston'. After a grand tour of Europe, he joined his father's textile business before being seized by a desire to preserve traditional Japanese culture after a voyage to the Far East. After ten years, he tired of this project and, on returning to Boston in 1893, he was given Camille Flammarion's book *La Planète Mars et ses conditions d'habitabilité*. He read it and scrawled 'Hurry' across his copy. Flammarion accepted Schiaparelli's theory of a canal network, rejecting competing theories that they were crevasses in ice fields or surface cracks caused by the cooling of the planet. They were, simply, too straight; they must be intelligently designed and constructed watercourses.

The book gave Lowell his new project. He linked up with William H. Pickering, who had been seeking funding for a Mars observatory in Arizona. Lowell raced out to Arizona with two telescopes, arriving at Flagstaff in March 1894. By June he had recorded a sighting of the Martian canal Lethes. A vast canal network was subsequently mapped. Lowell was convinced that its purpose was to irrigate the desiccated planet by bringing water from the polar ice caps. At once he launched a campaign of popular articles telling the world that there was – or had been – intelligent life on Mars.

His articles caused a sensation. Now famous, Lowell returned to Europe where, in Paris, he met Flammarion. They discussed not only Mars, but another mutual interest, the occult. In Milan he met Schiaparelli, who subsequently confided to a friend that Lowell needed to rein in his imagination. The eyes of the great Italian were, by this time, failing. In 1898 he was to retire from observational work.

Lowell was to produce three books about life on Mars and to fill lecture theatres wherever he went. His observatory was finally to map 183 canals, four times as many as Schiaparelli. But though he had persuaded the public, he had not persuaded the scientists. His theories were widely disputed. They were not, however, finally to be refuted until the Greek-born astronomer Eugene M. Antoniadi looked through a 33-inch refractor at Meudon. The canals, he saw, were a trick of the light, an optical illusion. Inferior telescopes had resolved random surface features into Lowell's mighty engineering works.

Just as Copernicus and Galileo had woken us from our dream of being at the centre of creation, so Antoniadi woke us from our dream of an advanced civilization on Mars. The structure of disillusionment is consistent down the ages.

But the idea of the Martian had been fixed in the twentieth-century imagination. He was the first alien of choice for the dreamers of science fiction dreams. Hugo Gernsback in his novel *Ralph 124C 41+: A Romance of the Year 2660*, serialized in his magazine *Modern Electrics* in 1911 and published in book form in 1925, naturally assumed that mournful, alien Martians would be part of our distant future.

'It was impossible to mistake the distinctly Martian cast of countenance. The great black horse eyes in the long melancholy face, the elongated slightly pointed ears were proof enough. Martians in New York were not sufficiently rare to excite any particular comment. Many made that city their permanent home, although the law on the planet Earth, as well as on Mars, which forbade the intermarriage of Martians and Terrestrials, kept them from flocking earthwards in any great numbers.'

Intermarriage may have been forbidden, but this particular Martian, Llysanorh, appears in the book as a sexual threat. Having cast his great black horse eyes on her, he aspires to abduct the lover of Ralph, the superintelligent hero. Aliens frequently try to carry off our women and delude our men.

Most vividly, there were the invading Martians in Wells' *The War of the Worlds* (1898) with their 'intellects vast and cool and unsympathetic, regarding this earth with envious eyes . . .' The panic caused by the invasion is, for Wells, caused by the horror of finding ourselves helpless before these monstrous creatures and their machines. The invasion delivers the message of modernity, that we don't amount to very much. We are reduced to little dots of lonely terror.

'If one could have hung, that June morning, in a balloon in the blazing blue above London, every northward and eastward road running out of the tangled maze of streets would have seemed stippled black with the streaming fugitives, each dot a human agony of terror and physical distress.'

That panic was mirrored in the real world in 1938. On the eastern seaboard of the US people fled their homes when *The War of the Worlds* was broadcast as a radio drama. Orson Welles' production used the conventions of news broadcasting to give the impression a Martian invasion was really happening. What is remarkable about this is that the people that fled New York in terror, in spite of the scientific evidence to the contrary, were prepared to believe at once that a civilization existed on Mars sufficiently advanced to mount such an invasion. Probably, in their imaginations, Mars was exactly where such creatures would come from.

In this context, even more strange was Dr J. E. Lipp's appendix to the Project Sign report in 1949, in which he mused on the location of the five nuclear explosions that had by then happened on Earth – 'Of these, the first two were in positions to be seen from Mars, the third was very doubtful (at the edge of Earth's disk in daylight) and the last two were on the wrong side of the Earth.'

Mars could still, at that time, have been considered as a possible site for an alien observation post or, indeed, as the home of a civilization, a civilization that, in Lipp's analysis, might be watching us. Our nuclear detonations had alarmed them and, as a result, their flying saucers had been sent to see what was going on.

Again there is this feeling of a visual one-way street. The Martians were seeing us clearly enough, but we saw them only fleetingly. It is we, not they, who have the optical illusions or suffer from tricks of the light. In the depths of the machine or somewhere out there in the hemisphere of the heavens, there are eyes that see clearly, alien eyes that penetrate without being penetrated.

'He realized that the glasses were surgically inset, sealing her sockets. The silver lenses seemed to grow from smooth pale skin above her cheekbones, framed by dark hair cut in a rough shag.'

This is Case, the hero of William Gibson's *Neuromancer* (1983), contemplating the impenetrable eyes of Molly, a seductive, optically cyborged warrior. Surgically implanted lenses are a challenge to the seducer. He cannot see her eyes so he cannot be sure he has her in his power.

Palmer Eldritch in Philip K. Dick's *The Three Stigmata of Palmer Eldritch* (1964) has had his eyes replaced by 'wide-angle luxvids'. These do away with eyeballs, replacing them with a horizontal slot that provides an exceptionally wide field of vision. Eldritch's gaze turns out to be multiply alien. His eyes are artificial, he inhabits the virtual world into which the hero, Mayerson, has plunged through the use of the drug Chew-Z in such a way that he can become any of the people Mayerson encounters, including his lover and, finally, Eldritch himself turns out to be an alien, an illegal who has smuggled himself into the solar system from 'Prox'. And so, in Eldritch's eyes, Mayerson sees the infinite regression of an alien eternity.

'And when he looked up into her face he saw the hollowness, the emptiness as vast as the intersystem space out of which Eldritch had emerged. The dead eyes, filled with space beyond the known, visited worlds.'

These are dangerous eyes. They seem to embody the empty hemisphere of space, a spectacle that, for Mayerson, is dangerous to contemplate. But alien eyes are often dangerous for more literal reasons. The robot Gort in *The Day the Earth Stood Still* has a metal visor where his eyes should be. He raises it not to see but to expose his weapon, a beam that heats, melts and destroys. Gort has a weapon instead of eyes. But it is a weapon of light, an optical weapon. Our eyes absorb light, Gort's project light. Again the alien gaze penetrates and, again, remains impenetrable – Gort is all smooth, featureless metal. Similarly, the eyes of the destructive robot in *Saturn 3* (Stanley Donen, 1980) are glass stalks with glowing tips. These are supported by a gantry which should, anatomically, be a neck supporting a head. But the 'head' is just eyes.

The eyes of the Borg, the ultimate villains in *Star Trek*, are as empty as those of Palmer Eldritch. But the emptiness behind them is not that of space but of the collective. There is no such thing as an individual Borg for this is a species that assimilates life forms it encounters into a single consciousness. When one Borg looks at you it is with the eyes of billions. And the Borg are, of course, cyborgs. Individuals are assimilated by a nanotechnology and by implants, one of which replaces an eye. The Borg gaze is highly selective. It will ignore you unless you are a direct threat and so the *Enterprise* crew can walk safely among the Borg unless they choose to do something aggressive. The implanted eye also emits a red laser beam – again these alien eyes project – so that a group of Borg create a dancing, linear light show.

Always the eyes are what we first see. Alien monsters, after all, are traditionally described as being 'bug-eyed'. In September 1952 in the town of Flatwoods, West Virginia, some children saw what they thought was a meteor land on a nearby hill. On their way to the hill to investigate, they were joined by Kathleen Hill, her two sons and a National Guardsman. When they got to the hill, one of the group saw what he took to be the eyes of an animal in the branches of a tree. A torch was shone in that direction and a huge figure was revealed. It was between ten and fifteen feet tall and had a blood-red face and glowing, greenish-orange eyes. The monster floated towards the group, all of whom ran back down the hillside.

When the eyes are wrong, everything is wrong. 'What have you done to his eyes, you maniacs?' screams Rosemary Woodhouse when she first sees her child, fathered by Satan, at the end of *Rosemary's Baby* (Roman Polanski, 1968). We are not shown the eyes, we are left to imagine the worst. But perhaps it is simply that they are uncanny, almost but not quite right.

There is a rare delusional disorder that sometimes occurs in schizophrenics or as a result of brain damage known as Capgras syndrome. It is a medical realization of the uncanny. Sufferers believe that somebody they know, usually a key figure in their life like a husband or wife, has been replaced by an imposter, robot or alien. The person looks no different, which explains why others cannot see what has happened. But the patient knows that something is not quite right.

This idea of an alien who appears, in all outward respects, to be a perfectly normal human being is not restricted to Capgras patients. It is a commonplace of alien lore. After all, if aliens can cross interstellar space, defy gravity and read minds, then they might well be able to adopt the camouflage of a human body. Georgina Bruni, author of *You Can't Tell the People*, about the alien encounter in Rendlesham Forest, tells me that she is convinced that a man named Alan, whom she knew for fourteen years, was from another world. He made spectacularly accurate predictions and delivered strange pronouncements. He died in his seventies, but his hands were always those of a twenty year old.

'I don't know who he was from that day to this,' she says. 'I've seen him do the most incredible things.'

Alan was benign, but the idea of alien seizure of human form is more commonly horrific. In *Invasion of the Body Snatchers* (Don Siegel, 1955) alien invaders leave behind pods inside which replicas of nearby humans are grown. When they are fully formed, they conquer. The plan is to use this method to take over the entire world. The film is about how easily our eyes can deceive us and how hard it is to accept that we are being deceived. It is set in the suburban surroundings of the super-normal Californian town of Santa Mira. Those taken over continue to behave plausibly. But they have no emotions. And it is this, rather than anything about their appearance, that gives them away. In the climactic scene, the hero realizes his girlfriend has been replaced with a replica only when he kisses her and feels nothing in her lips. As he draws back, her eyes have become alien and cold.

But the gaze of alien visitors is not always so frightening. In October 1954 at Ranton, near Shrewsbury, Mrs Jennie

Roestenberg said that she and her two children had watched a disc-shaped object hover over their house. There were two transparent panels on the side through which she could see two creatures with white skins, long hair and high foreheads. They had transparent helmets and turquoise clothing. The disc was tilted and the two creatures looked down 'sternly, not in an unkindly fashion, but almost sadly, compassionately'.

There is pity in this alien gaze. We are not worthy. We are, in fact, revolting. Perhaps he was a keeper in the zoo.

Gerald Heard, a friend of Aldous Huxley, published *Is Another World Watching? The Riddle of the Flying Saucers* in 1951. It was written at the height of the first postwar UFO wave. Heard believed that the alien visitors were benign and took it for granted that they came from Mars where 'life is certainly far ahead of us'. At the time it was thought that no saucers had landed, so the appearance of the aliens was unknown. But, taking his cue from speculation that the Martians were likely to be intelligent insects, he said the pilots of the saucers were super-bees 'of perhaps two inches in length . . . as beautiful as the most beautiful of any flower, any beetle, moth or butterfly. A creature with eyes like brilliant cut-diamonds, with a head of sapphire, a thorax of emerald, an abdomen of ruby, wings like opal, legs like topaz – such a body would be worthy of this "supermind" . . . It is we who would feel shabby and ashamed, and maybe with our clammy, putty-coloured bodies, repulsive!'

We shrivel with shame under the alien eyes. Their gaze is too much to take. As John Mack writes of the typical experiences of the hundreds of alien abductees he has interviewed – 'The eyes . . . Have a compelling power, and the abductees will often wish to avoid looking directly into them because of the overwhelming dread of their own sense of self, or loss of will, that occurs when they do so.' 'I see their eyes,' says Mack's abductee Sheila, 'I don't want to see them anymore.' 'The big eyes are scary, Mommy,' says child abductee Colin. 'Their eyes. I just hate 'em. I hate 'em,' says Jerry. 'It's like they're looking right through you . . . They go inside you . . . a really weird, unnerving feeling . . . It's as if I'd lose my self, and don't feel like I have any control.'

In this context, it is a strange accident that the first mainstream report of an alien abduction – that of Betty and Barney Hill – appeared in 1966 in *Look* magazine. For the power of the aliens is in their look. Their eyes are like weapons, but not necessarily of destruction like Gort's beam. They may be weapons of control: they control by seeing through humans, but the humans cannot see through them. The aliens, therefore, control light and light, we sense, is at the very centre of reality.

'It is here,' writes Mack, 'perhaps, that the physical and non-physical meet, where spirit and matter intertwine, and where the unknown and the unknowable (at least by the methods of empirical science) touch each other. For light quanta, photons, are without mass, and light appears to exist everywhere with no apparent physical substrate.'

Mack's abductees typically go through darkness into light, an image that occurs repeatedly in dreams and the occult as a sign of movement beyond death. Abductee Julie sees herself as 'an eternal body of light'. Nona had an out-of-body experience in which she feels the aliens were deliberately showing her that we are beings of 'glowing light' or a 'light force'. Aliens in films frequently emerge from blinding white light – most notably at the end of *Close Encounters of the Third Kind* (Steven Spielberg, 1977) – indicating their habitat is the beyond, beyond the stars, beyond death.

The eyes have always been the primary sense organs, the windows of the soul. The idea of the window suggests they do not only look out but can also be seen through. We look at other eyes that see. This pre-modern idea has been endorsed by the modern sense of light as an absolute. One of the great – and certainly one of the most popularly accepted – concepts of twentieth-century physics was that the speed of light was, indeed, an absolute. It is the ultimate speed. Nothing can travel faster, not even two photons moving away from each other at light speed. Their speed is not twice that of light, it is only that of light. And so light becomes the absolute on which the universe is constructed and inside which it must exist. Or, perhaps, only our universe. One of the commonest accomplishments of aliens is an ability to travel faster than light. They have relativized our absolute. One investigator – John Keel – thought the aliens were actually composed of light.

The physics thus endorses the magical sense that the eyes are the focal point of our anatomy and that the acts of seeing and being seen are almost unbearably attuned to our deepest nature. And the immateriality of light, noted by Mack, further informs the sense that the eyes and sight are close to spirit and soul. Eyes are our contact with the ultimate.

But twentieth-century physics has also put a limit on what we can know through light. Heisenberg's uncertainty principle traps us in the act of looking. It states that, at the quantum level, we cannot know everything about a particle – position, velocity, direction. Observing one denies us knowledge of another. Observation changes the thing observed. And Major Angus G. Burnside in H. Chandler Elliott's story 'Inanimate Objection' (1954) knows that, in response, things rebel.

'Well, you know that the presence of an observer changes conditions so you can't know what would have happened with no observer. That won't perceptibly affect motion of a falling body, or other such elementary cases. But in complex, versatile systems, I believe the effect increases enormously. I believe physical processes *know* they're being observed and evade analysis – I'm using "know" as engineers do when they use the expression, "How does the valve know force is applied?"'

Inanimate objects in the story respond to man's observations and meddling by fighting back. They resist our attempts to organize the world. They cause natural disasters to occur. This is not a capricious rebellion of the inanimate world, it is built into the fabric of the cosmos and it means no aliens will ever visit us. For, long before a civilization can achieve the technology of interstellar travel, matter, incensed at being observed, will bring it down. In this case, the aliens really are all around us, they are all inanimate matter that, unknown to us, burns with resentment. Only Major Burnside knows this.

There are other limits, apart from uncertainty, on what we can know through sight – convention, for example. We do not see because we are conditioned to be blind to anything that threatens the prevailing world view. This was the obsession of Charles Fort, one of the great seers.

'Our position,' he wrote, 'That the things have been seen: Also that their shadows have been seen.'

Born in Albany, New York, in 1874, Fort is a mighty figure in the realm of the astounding. He was also physically mighty, his friend Theodore Dreiser likened him to Oliver Hardy. Fort created a tradition, still sustained by his disciples, of collecting strange and always inconclusive tales that demonstrated that the world of conventional science, common sense and quotidian reality was a mere surface phenomenon concealing the wonders beneath and beyond.

'Enormities and preposterousnesses will march,' was his battle cry.

His central work in the context of aliens is *The Book of the Damned*, published in 1919. The Damned are all those areas of knowledge excluded by conventional understanding. The book is an astonishingly prescient catalogue of alien lore. Fort anticipates Erich von Däniken's theories about aliens landing on Earth thousands of years ago. He speculates, as many have since done, that the human species is some kind of experiment – 'I think we're property. I should say we belong to something' – and he advances the theory of panspermia, now embraced by many respectable scientists, that life spreads through the universe and humans were seeded from space. Fort's point was not that any of the wonders and possibilities he catalogued were necessarily true, but that they lay all around us uninvestigated because of the deadly grip of conventional wisdom on the human imagination.

We see conventionally. It is not only that we think and act and speak and dress alike, because of our surrender to social attempts at Entity, in which we are only super-cellular. We see that it is *proper* that we should see. It is orthodox enough to say that a horse is not a horse, to an infant – any more than is an orange an orange to the unsophisticated. It's interesting to walk along a street sometimes and look at things and wonder what they'd look like, if we hadn't been taught to see horses and trees and

houses as horses and trees and houses. I think that to super-
sight they are local stresses merging indistinguishably into
one another, in an all-inclusive nexus.

Whole worlds and vast artefacts that Fort calls 'super-
constructions' seem to drift by us unnoticed.

'I think that it would be credible enough to say that many
times have Monstrator and Elvera and Azuria,' Fort writes of
some of these unseen planets, 'crossed telescopic fields of vision,
and were not even seen – because it wouldn't be proper to see
them; it wouldn't be respectable, and it wouldn't be respectful: it
would be insulting to old bones to see them: and it would bring
on evil influences from the relics of St Isaac to see them.'

And the super-constructions seem to be huge and wheel-like.
Fort thinks they are made to 'roll through the gelatinous medium
from planet to planet'. Sometimes they enter Earth's atmosphere,
a less dense medium that may cause them to explode. To avoid
such a catastrophe, they must submerge themselves in the sea.
Seamen have seen them not far from the Persian Gulf.

Fort is the most extravagant exponent of the view that we have
been literally blinded to reality by modern science. We are like the
complacent townspeople of Santa Mira, we cannot see what lies
beneath the quotidian. Science allows only certain types of
knowledge, types that accord with accepted theory. The bleak
emptiness that we see when we look up at the hemisphere of the
heavens is a delusion. In reality, it is full of life and wonders.

Or life offers a fulfilment we have denied ourselves by closing
our eyes to the cosmos. Credo Mutwa, an African medicine man,
told John Mack: 'The entire Western civilization is based upon a
blatant lie, the lie that we human beings are the cocks of the walk
in this world, that we are alone and that beyond us there is
nothing.'

We are fallen into blindness. Barbara Marciniak channels
messages from beings in the Pleiades star cluster, 425 light years
away. They tell us we must free ourselves, we must see.

'When the ancient eyes are open,' the Pleiadians say, 'and you
recognize your true potential, you will stop arguing with
yourselves.'

Sir John Whitmore, a businessman and former racing car driver, decided in his thirties that there must be more to life than getting and spending. He came across a group that has, in the past, included the singer John Denver and Gene Roddenberry, creator of *Star Trek*. Via the medium, this group is in touch with the Nine, a committee of cosmic beings. The Nine wish to communicate the true nature of the cosmos, a nature we cannot see. Specifically, we cannot see that there are many interlinked universes.

'The Nine,' Sir John explained to me, 'talk about the structure of the universe . . . The big problem we have with the understanding of this whole field is that limited belief in a physical universe that operates in one time-space envelope . . . if that's the frame you look at it in, doesn't even include the possibility that there are other interlocking universes.'

All of these voices tell us that we are cursed by delusions of objectivity. We think that what we see with our modern eyes is all there is to be seen. Or, as Patrick Harpur says, it is 'a purely Western characteristic to confuse the literal and the physical'. There is a world invisible to literal eyes, a world routinely seen by people before science blindfolded us. It is the world, for Harpur, of daimons – 'We need double vision to see daimons – to see that they are real, but not literally so.' The modern daimon is the alien.

But Western objectivity is more obviously perverse than this. For, having shown how small we are in the universe and how contingent, it then insists that, nevertheless, our eyes are uniquely accurate organs. They and only they can apprehend the real. This is self-evidently not true. To a neutrino I and the desk at which I sit are an all but empty haze. Is that not as real as the solidity I perceive? The very existence of truly alien eyes subverts the vanity of our science, reducing it to one perspective among millions.

And yet we still long for visual proof that we are not alone; the intuitions, the tales, the voices are not enough. In the cramped offices of *UFO Magazine* in Leeds, the editor Graham Birdsall showed me videos of ambiguous objects flying through the air. In Roswell, New Mexico, you can visit museums and see artefacts of the 1947 saucer crash and buy earth from the crash site.

Ufologists search desperately for the visible evidence that the aliens have landed or crashed and insist that governments have concealed real, sensational evidence for fear of public panic or to keep the technology to themselves. We may say our eyes are blinded by narrow-minded contemporary science, but we still want to know for sure, to see with those eyes, to see eyes that look back, penetrating but impenetrable from the heavens.

But, in the end, like Krikkit and Wesker's World and in spite of Galileo, our planet remains shrouded by an opaque cloud. We cannot, after all, see out and neither can we see the cloud, for it is an imagined one of our own making. Trader Garth should come crashing through any day now. We will stare at him and he will stare back.

13

Enrico Fermi

In 1950 Enrico Fermi, Nobel prizewinner, creator of the world's first nuclear reactor and professor at the Institute of Nuclear Studies at the University of Chicago, was at lunch with some colleagues in Los Alamos, New Mexico, the place where the atom bomb had been developed. The conversation turned to aliens and the possibility of an inhabited universe. It was a predictable topic in view of the fact that the United States and, increasingly, the world was then being buzzed by flying saucers. But Fermi was no believer.

'If there are extraterrestrials,' he asked, 'where are they?'

Fermi's authority was immense and this may have blinded people to the fact that the question had been asked many times before, though perhaps by less respected observers. Most notably there was Charles Fort, who pondered the matter in his *Book of the Damned* of 1919.

'The greatest of mysteries:

'Why don't they ever come here, or send here, openly?'

He concluded: 'It's probably for moral reasons that they stay away – but even so, there must be some degraded ones among them.'

But the real point about Fermi's question – which was to

become known as the Fermi paradox – is that it has some surprising scientific depths. The usual debate about the possible existence of extraterrestrial intelligence is familiar. Believers say we have been visited by aliens but they have concealed themselves from us. Sceptics say the believers are deluded and the conceal-ment theory is just a convenient device for covering up the complete lack of physical evidence. Sceptics, however, may accept the possibility that there are aliens somewhere out there in the immensity of space. Here they might find themselves agreeing with believers on the basis that it seems improbable in such a vast universe that we should be the only intelligent beings. The core of the sceptical position is, therefore, simply the insistence that all these reports of alien visitations are false. But their falsity does not preclude the possibility that aliens exist. The correct answer to Fermi's question should be, in this interpretation, 'If anywhere, elsewhere.' But, perhaps simply because it was Fermi who asked, people soon began to work out that there was a good deal more to the question than met the eye.

Assume, for the moment, that all those who have seen UFOs, encountered aliens or been abducted by them are mad or deluded. Assume, in fact, that there is absolutely no evidence of intelligent life elsewhere in the universe.

This might mean that the aliens have not managed to make contact because of the distances involved. Even at close to light speed, it might be too far to come or call. We know what an effort it has been for humans to make the short jump to the moon and even our most venerable deep space probes have not, in galactic terms, yet left our neighbourhood. So it seems to be simple common sense to say that distance keeps the cosmic community apart.

But, in fact, this might be a very poor argument indeed. Say technologically advanced civilizations could produce spaceships capable of one tenth the speed of light – 30,000 kilometres per second. This is a reasonable assumption. Higher speeds might involve an impossibly high energy requirement. At that speed it would take them only one million years to populate the entire galaxy. This is not long in a universe that may be 15 billion years old. But, plainly, aliens have not colonized the galaxy. They are not on Earth and our radio telescopes have not picked up any signals. They are neither here nor there.

This argument – essentially a detailed footnote to Fermi's paradox – was first advanced in 1975 and took on enormous force in academic circles. It had a pleasing combination of clarity and statistical weight. The universe was big, it said, but not too big. The question 'Where are they?', therefore, was a serious blow to the believers' case because, if they existed, they were likely to be here.

Previously, the most respectable way to calculate the probability of intelligent extraterrestrial life had been the Drake equation. This had been formulated in the early sixties by the astronomer Frank Drake, the primary inspiration behind SETI, the search for extraterrestrial intelligence. The equation was based on a series of key variables: the rate of formation of stars in the galaxy, the fraction of those stars with planets, the number of planets per star capable of sustaining life, the fraction of planets where life evolves, the fraction where intelligent life evolves and the period of the planet's life in which a communicating civilization endures. This should produce N, the current number of communicating civilizations in our galaxy.

The Drake equation has a reassuringly scientific air, but it is, in fact, only a sophisticated mathematical expression of the intuitive sense that because the universe is very big, therefore there must be intelligent life elsewhere. It puts this in practical terms by insisting that such life must be communicative. If it were not – either because it was so advanced it had decided there was no point or so backward that it had not found out how to manipulate the electromagnetic spectrum – there would be no way we would ever know of its existence.

The Drake equation became the key rationale for SETI, a radio survey of the heavens seeking signs of a communicative civilization. SETI was run by NASA until 1994 when congressional scepticism drove it into the arms of the private sector. NASA's more recent interest in astrobiology is a slightly more politically respectable way of considering the possibility of extraterrestrial life because it is cheaper and does not involve what had come to be seen as a rather gormless, if not downright geeky, scanning of the heavens.

SETI, privately or publicly funded, has found nothing, though there have been several 'Wow!' moments – so-called because that's what the observer said when the first one happened. In 1960, long before SETI was established, Drake's first survey picked up a highly coherent signal that turned out to be from a passing aircraft. There have been many other such false alarms, the most thrilling of which was perhaps the signal picked up by Jocelyn Bell, a Cambridge graduate student. This was an absolutely regular radio pulse. It must be artificial. It wasn't. It was produced by a pulsar, a spinning pair of neutron stars, the remnants of supernovas. In honour of this Wow! moment, the first pulsar was named LGM1, the letters standing for Little Green Men. Ufologists would, of course, have told the scientists that the 'G' should have stood for grey.

But the mathematical justification of the Fermi paradox had cast doubt on the whole SETI idea. The implication was that it was pointless scanning the stars because the mere fact that the aliens weren't here was proof enough that they didn't exist. When combined with the fact that the scans have so far produced nothing – a silence now known resonantly as the Great Silence – the case for our utter solitude in the immensities of the Milky Way seems overwhelming.

As the philosopher David Lamb points out, there is an uncanny parallel between this debate and the theological issue of the *silentium Dei* – the silence of God. If God exists, why does He keep so quiet about it? Bertrand Russell was once asked what his answer would be if God ever challenged him on his atheism. Russell replied he would ask God why He had offered so little evidence for His existence. Aliens, if they exist, are similarly evasive, both in terms of the Great Silence and in the way that those who have arrived have apparently gone to such elaborate lengths to conceal themselves.

Various arguments have been advanced to account for the Great Silence. There may be a fundamental problem with space travel. Aliens may not be interested in communicating. Maybe they haven't had enough time to reach us. Or, possibly, they have visited us and we are not aware of it. Von Däniken's ancient astronauts would plainly be covered by this last point. All of these are underpinned by the conviction that, as there are, according to some estimates, 100,000,000,000,000,000 Earth-like planets in the universe, there simply must be somebody else out there.

These are all, of course, speculative points, as are the obvious refutations, as, indeed, are all the possible solutions of the Drake equations. Most of the variables were unknown and could only be estimated by wild guesswork. But the structure of the dispute, as Lamb implies, is significant. For this structure defines the intellectual parameters of the emotional reaction to the spectacle of the heavens at night.

The strength of this reaction has always harassed the Western imagination. Aristotle spoke of it as wonder: 'It is owing to their wonder that men first began to philosophize; they wondered originally at the obvious, and then they advanced little by little to the greater matters, about the phenomena of the moon and those of the sun and the stars.'

Kant saw the spectacle of the night sky as one of the two great miracles of creation: 'Two things fill the mind with ever new and increasing admiration and awe: the starry heavens above and the moral law within.'

Even Enrico Fermi was struck by a remark of a farmer he overheard one starry night – 'What a nice sky! How can some people say that God does not exist?' There seems to be, as George V. Coyne of the Vatican Observatory remarks, an 'almost inevitable link in people's minds between the exploration of the Universe and thoughts about God'.

The first link is, of course, visual. The astounding purity of the blackness of the unpolluted night sky and the gleaming brightness of the stars is in sharp contrast to the chaos of our daytime visual field. It is unsurprising that the geocentric universe was founded on the belief that all matter beyond the moon was in a state of purity and perfection and even less surprising that this purity inspired thoughts of a Great Designer.

But the second link is distance. Even the Ptolemaic universe was large, but the Copernican universe is vast beyond all imagining. And, because it is beyond all imagining, there is an urge to populate it, if not with God, then with aliens. The question 'What is all this emptiness for?' may be irrational to the scientific mind, but it has always sprung naturally to human lips. Even the hard-headed scientists who insist that the cosmos is, indeed, empty and pointless are only saying this because of the strength of the opposing reaction. They are like the atheists, that strange cult that seems to require the existence of God more than anybody else.

From Epicurus onwards the idea of useless space has been an affront to the human imagination. It is, perhaps, the example of the Earth, every corner teeming with life, that persuades us that space must be filled with something other than a vacuum and rocks. It can be filled in two ways – either with life or with purpose. If with life, then we have either yet to find it or it conceals itself from us with fantastic ingenuity. If with purpose, then . . . well, we have tried through metaphysics and, latterly, physics which now tells us the universe might have to be this big just to produce us.

The blackness of the heavens at night is, we now know, distance, inconceivable distance. Distance is what grips the modern imagination when confronted by the starry heavens above, distance and the need to believe there is meaning in all those empty years of light. Space is the beginning and end of faith.

14

Alien Distance

'Our number came up in the Monte-Carlo game,' wrote Jacques Monod. 'Is it surprising that, like the person who has just made a million at the casino, we should feel strange and a little unreal?' Conscious life emerges by chance. And then, precisely because this emergence is so improbable, finds itself alone in the cosmos.

'The eternal silence of these infinite spaces frightens me,' confessed Blaise Pascal.

Carl Gustav Jung wrote of the human unconscious that it 'strives to fill the illimitable emptiness of space'.

'The more the universe seems comprehensible,' the physicist Steven Weinberg has said, 'the more it also seems pointless.'

'Space,' says Captain James T. Kirk, 'the final frontier.'

One way of describing the modern condition is to say that we are lost in space and time. For most of human history both space and time were theoretically navigable. The stars were simply overhead, a long way off but not *that* far, and the world was made a few thousand years ago. Even more consoling, humans were thought to be involved in both the contents of space and the passage of time. God created the universe in, according to the estimate of Bishop James Ussher in the seventeenth century, 4004

BC. And the cosmology generally accepted until the era of Copernicus and Galileo stated that He placed the Earth at the centre. Our free will and subsequent journey to salvation or damnation were the whole point of the exercise. Other faiths had other views, of course, but the Western Christian view was unique in one key respect. It was to be utterly subverted from within. Neither faith nor reason could have anticipated the severity of the temporal and spatial vertigo that was to afflict the Western imagination with the advent of modern science.

The first shock was the abyss of time. Eighteenth- and nineteenth-century geology established that the Earth was in constant motion. But this motion was very slow. Bishop Ussher's schedule did not allow enough time for the creation of the geological formations we see around us today. The summit of Everest, we now know, is made of marine limestone; what incalculable eons had to pass for that to be raised from the depths of the ocean? Millions more years were required for this slow, mineral dance to take its course. In fact, according to our current estimates, it has been 4 billion years since the planet was formed. At once, the whole human narrative plummets, unnoticed, into the abyss of deep time.

One Victorian artist, William Dyce, captured the mournful mood induced by this realization in his painting *Pegwell Bay: A Recollection of October 5, 1858*. There is a rock-strewn seashore overlooked by low, dark cliffs. There are people on the beach in the foreground and more in the distance as well as some horses. There are a few strands of cloud in the evening sky. There is an, at first, indefinable air of melancholy. Slowly it dawns on you what it is. The people seem transient, merely walk-on extras in this scene. They are less real than the ancient rocks in the background. They are just passing through, brief flashes of consciousness against the imperceptibly slow upheavals of the rocks. The people had become alien visitors in their own world. The painting is suffused with the peculiar melancholy that haunted the Victorian mind.

Worse was to come. The second law of thermodynamics revealed that all systems run downhill. A hot drink dissipates its heat into the cooler air of the room. There is no appeal against this law. In time all possible reactions in the universe will cease as energy levels equalize and the whole thing slumps into 'heat death', a terminal, warm inertia. Not only was the human story too vanishingly small to be of any significance, it was also doomed to absolute extinction. No memory of our tribulations nor fragment of our achievements would remain.

With the twentieth century came the abyss of space. We always knew it was big, we just hadn't realized it was this big. It was as big as time, in fact. We learned to use light years – the distance light travels in a year – as a measurement of distance and to view the size of the universe as a function of the time elapsed since it came into existence. And that time was around 15 billion years. The abyss of time and the abyss of space were the same thing and in human terms, they were both unimaginable. The light year is not a measure of distance, it is the index of our loss.

But maybe there is a sliver of hope. Maybe this spatio-temporal abyss is, in fact, populated. The hard sceptic will deny the reality of UFOs or claims of contact or abduction. But that is only to say that aliens have not visited Earth. What even he cannot honestly deny is that the universe is very big. Simple common sense would suggest that out there, somewhere, are other intelligent beings. If there are not, then these vast spaces are, indeed, eternally silent, and illimitably empty.

They would also seem to be offensively pointless. Emptiness on such a scale is an affront, an implausibility. The human mind is not inclined to accept this possibility. And not just the modern mind – since the time of Epicurus (341–271 BC) it has been argued that there is, within, nature, a 'principle of plenitude'. This states that no potentiality of being can remain unfulfilled, that there is an inexhaustible source of being in nature – later identified with the Christian God – which will ensure that the universe is, in fact, full of countless forms of life, solely because such forms are possible.

This tradition ran counter to the Aristotelian–Thomist orthodoxy that we were at the centre of a cosmos designed with us in mind. In fact, frequently it went so far as to suggest that, not only was the universe populated, it was also populated by superior beings. The fifteenth-century scholar Nicholas of Cusa followed Epicurus on the subject of many worlds and went on to speculate about the nature of aliens.

> Rather than think that so many stars and parts of the heavens are uninhabited and this Earth alone of ours is peopled – and that with beings, perhaps of an inferior type – we will suppose that in every region there are inhabitants, differing in nature by rank and all owing their origin to God, who is the centre of circumference of all stellar regions . . . It might be conjectured that in the area of the sun there exist solar beings, bright and enlightened denizens, and by nature more spiritual than such as may inhabit the moon – who are possibly lunatics – whilst those on earth are more gross and material.

The astronomer Johannes Kepler (1571–1630) asserted that the moon was occupied by strange, wandering beings.

'In the course of one of their days, they roam in crowds over their whole sphere, each according to his own nature: some use their legs, which far surpass those of our camels; some resort to wings; and some follow the receding water in boats; or if a delay of several more days is necessary, then they crawl into caves.'

Richard Bentley, a seventeenth-century English divine, and one of the first to recognize the genius of Isaac Newton, was puritanically opposed to the Epicurean view of a cosmic cornucopia of life. But, equally, he could not bear the thought that the stars just existed.

'All Bodies,' he wrote, 'were formed for the Sake of Intelligent Minds: As the Earth was principally designed for the Being and Service and Contemplation of men; why may not all other Planets be created for the like uses, each for their own inhabitants who have Life and Understanding.'

The greatest philosopher of the Enlightenment, Immanuel Kant, in his *Universal Natural History* took it for granted that the stars were suns. He thought the cosmos was 'animated with worlds without number and without end'. These worlds were inhabited by rational beings that grew ever more refined the further they were from the centre.

This pluralist impulse to populate the universe, to give it some purpose, to save it from utter pointlessness is, as historian Michael J. Crowe points out, a human constant. Seven hundred years ago, for example, the medieval scholar Albertus Magnus described the issue of plurality as 'one of the most wondrous and noble questions in Nature'. Humans have always been engaged in the quest for the other to fill the emptiness, to ease the burden of their loneliness.

For some, this is an urgent issue, a matter of life and death. Pluralism, like religion, is reluctant to allow itself to be undermined by contrary evidence and will constantly seek ways to preserve its faith.

'The generalized pluralist position,' writes Crowe, 'is clearly unfalsifiable. If the moon be found bereft of life, Mars remains. If Mars be barren, Venus is available. If the solar system is shown to lack life except on Earth, emphasis can be placed on planets orbiting other stars . . . Moreover, the assumption of an omnipotent God unconstrained by laws of terrestrial life expanded almost endlessly the repertoire of rescue techniques available to pluralists.'

The point is, of course, that this defence of pluralism can go on for ever. However much we might explore, there is always so much more space where the aliens might still be waiting. There are approximately 100 billion stars in the universe for every human being on Earth. We do not know how many are orbited by planets, but, over the past decade, we have discovered that some definitely are. Given the limitations of our instruments in detecting objects in the vicinity of a star, this almost certainly means that there are many planets beyond our solar system, billions maybe, trillions probably. Is there life on any of them and, if so, is it intelligent?

Calculating the probability of extraterrestrial life is not easy since we have only one model – that of Earth – on which to base our estimates. The primary issue is: does life happen the moment it gets a chance – known as the assumption of necessity – or is life on Earth a single, wildly improbable occurrence? Scientists have always been and remain divided. Both sides appear to be able to argue from common sense principles.

Plainly, the Great Silence of the universe, in spite of the best efforts of our radio astronomers, would seem to support the wildly improbable hypothesis. Our number just happened to come up in the lottery. Furthermore, the vast temporal abyss should have given species more advanced than our own time to make their presence felt. Yet, as Enrico Fermi asked, where are they? Common sense tells us they are nowhere.

On the other hand, the fact that we are here does mean that life happened at least once. If Earth is the only planet on which this happened, then it is, indeed, a fantastically improbable event. Fantastically improbable events don't often happen. It is, therefore, likely that life happened elsewhere. Common sense tells us there are aliens out there.

The simple truth is that our narrow, earthbound perspective makes it extremely difficult to establish any clear idea at all about the likelihood of life elsewhere. Every hypothesis seems to melt away the moment it is formulated.

One celebrated false alarm, for example, was the Miller–Urey experiment in 1953, in which a simulation of the primordial soup of Earth was created and subjected to artificial bolts of lightning in a flask. After a few weeks amino acids – the constituents of proteins – were found in the soup. This, it was said, demonstrated the simple mechanism whereby life began. But, we now know, amino acids are 'thermodynamically downhill'. They are formed, in other words, very easily. Life is definitely 'thermodynamically uphill' and nothing like as simple.

Also, organic molecules have been detected in space. This has been used as evidence for the 'panspermia' hypothesis – essentially that life on Earth was 'seeded' from space. This was advanced most persuasively by Fred Hoyle and Chandra Wickramasinghe in their book *Lifecloud*:

> The best explanation therefore of the known facts relating to the origin of life on the Earth is that in the early days soft landings of comets brought about the spreading of water and other volatiles over the Earth's surface. Then about four billion years ago life also arrived from a life-bearing comet. By that time conditions on Earth had become

sufficiently similar to those on the cometary home for life to be able to persist here, probably at first tentatively and then with some assurance as time went on. The long evolution of life on Earth had begun.

We, then, are the alien invaders of Earth. This, of course, does not answer the question of the origin of life before it came here. But it does evoke an Epicurean sense of the pervasiveness of life. It is, in this view, a natural aspect of the entire cosmos. Spawning life is simply what matter does. Panspermia is the contemporary scientific version of the principle of plenitude. Nevertheless, to Hoyle and Wickramasinghe, it is an idea of the utmost modernity. The concealed leap of faith – that being cosmically insignificant must mean we are not alone – is interesting.

'The idea that in the whole universe life is unique to the Earth is essentially pre-Copernican. Experience has now repeatedly taught us that this type of thinking is very likely wrong. Why should our own infinitesimal niche in the universe be unique? Just as no one country has been the centre of the Earth, so the Earth is not the centre of the universe.'

There has also been the continuing debate about life on Mars. One experiment from a US Viking lander in the 1970s did seem to indicate the existence of microbial life on the planet. This was later discounted by NASA as arising from our ignorance of the constituents of the Martian atmosphere. In the great tradition of alien cover-up theories, however, Gil Levin, the scientist who designed the experiment, still insists it demonstrated the existence of life. In the late nineties there was also the Martian meteorite found in Antarctica. This appeared to contain bacteria-like fossilized structures. After an initial NASA announcement that these did seem to be evidence of Martian life, there has been a considerable climb-down. The recent US landers seemed to strengthen the case for microbial Martian life. But the evidence remains ambiguous. Confirmation that there was once water on Mars is not much of a development. It was always thought to be the case.

The extraterrestrial life issue in mainstream science remains, therefore, unresolved and apparently unresolvable. But, over the past thirty years, the scientific consensus on what the answer might be has changed. It has moved from 'probably no' to 'possibly yes'.

There are two primary reasons for this change. The first is the discovery of planets outside the solar system. In practice, this is not a very good reason because it was always overwhelmingly likely that such planets would exist; finding them merely removed the very remote possibility that there was only one planetary system – ours – in the entire universe.

The second reason is slightly better. This is the discovery of organisms called extremophiles – lovers of extreme conditions – on Earth. Strange worms have been found living around super-heated vents on the ocean floor. Also there are SLIMES (sub-surface lithoautotrophic microbial ecosystems), assemblies of bacteria that can live in rocks to depths of two miles or more because, unlike all other life, they do not need to consume organic particles; they eat, in effect, rock. If all life were extinguished on the surface, SLIMES would survive and they may, indeed, now be living deep beneath the surface of Mars. A bacterium has been discovered – *Deinococcus radiodurans* – that can survive doses of radiation of up to 3 million rads. A mere 1000 rads is enough to kill a human. It can survive in the reactors of nuclear power stations. It does so by constantly repairing its own DNA. We have no idea how.

Prior to the discovery of such creatures, it was thought that life could only exist in a very narrow range of environments. This was a demonstration of the strange persistence of the pre-Copernican view. Scientists may not have continued to believe the sun orbited the Earth, but they did suffer from another unsupported, anthropocentric delusion. Life, they assumed, must be somewhat like us and, therefore, must require conditions somewhat like the ones that sustain humans. Perhaps, in fairness, this is just a working assumption. If life is utterly different from us – like, for example, Douglas Adams' 'super-intelligent shade of the colour blue' – then we may be unable to detect it and, even if we did detect it, how could we possibly talk to it? Any attempt to find such creatures would have to be based on a definition of life of such generality that the search would be impossible. It would involve the interrogation of almost every object in the universe. Are you or are you not alive?

Extremophiles demonstrate, first, that life may well be rather more unlike us than we realized and, secondly, that there is a higher probability that, given the chance, it will get going simply because we now know there are more potentially life-sustaining environments.

Further support for this comes from fossil evidence that life began on Earth only about 200 million years after the planet was formed 4 billion years ago. This seems to indicate that life springs more or less automatically into existence when the conditions are right. If it were merely a freak occurrence, then 200 million years is not long enough for the necessary molecular shuffling to make it likely. Straightforward probability calculations of the chances of such shuffling producing life have suggested that it remains virtually impossible after billions of years. One estimate, based on the assumption that the entire planet was covered by an ocean ten kilometres deep, was that there was a one in a 100 billion chance of bacterial life arising over a period of 1 billion years.

However, though the scientific consensus has swung towards the view that there may be life out there, the change is tentative, a slight tilting of the default condition which is, of course, ignorance. Either to believe it is likely that there is life or to believe that it is not is a matter of subjective choice. Unless, of course, you have met it.

If you are not religious, which way you incline will depend on your view of the nature of matter. The classical view of Enlightenment science is, more or less, that it is just inert stuff that, for reasons we cannot understand, obeys certain rules. These rules do not impose any particular direction on matter. We are thus utterly contingent products of the operation of rules of great generality upon matter.

Darwinian evolution modifies this slightly. It says that once life gets started, there is some direction to the behaviour of matter. The strength of this directional push depends on your interpretation of Darwinism. Some would say that, once replicating molecules appear, the ultimate appearance of intelligence is more or less inevitable as it appears to be the most effective gambit in the game of life. Alternatively, you could argue that the directional push of evolution is minimal. All it does is select the best adapted organisms. But, since what is involved in this adaptation is dependent on a constantly changing environment, there can be no particular direction. Different organisms – the dinosaurs, us – will be successful at different times in different environmental conditions. Either way, Darwinism is strictly limited to the behaviour of matter once life has begun. It says nothing about how or why it begins. It does not, therefore, necessarily modify the fundamental view of our emergence from dumb matter as a staggering accident, as improbable as our number coming up at Jacques Monod's Monte-Carlo game.

Latterly, however, a view has emerged that there is, indeed, an overall direction to matter that would make life more probable. Essentially, matter inclines towards complexity. The working of the laws of physics intrinsically push matter in the direction of ever more complex forms and life is nothing if not complex.

'Life emerges,' says physicist Paul Davies, 'as a natural part of the outworking of the laws of physics and will occur wherever and when there are earthlike conditions.'

In this case, it is possible to say that it is in the nature of the universe to produce life and, therefore, there is likely to be a lot of it. Davies is an adherent of this view, though he admits this is largely a personal inclination. There is no strong evidence either way. But he is an optimist, believing that aliens exist and that contact with them will 'restore to human beings something of the dignity of which science has robbed them'.

Unfortunately, even if the Davies view is correct, this does not guarantee an alien encounter. There are many reasons to believe we may never know that we live in a populated cosmos.

Maybe, for example, the aliens are avoiding us simply because we are too dull or backward. The Vulcans in Star Trek did not make contact with humans until they noticed we had discovered warp drive. Ford Prefect's total entry on our planet in Douglas Adams' Hitchhiker's Guide to the Galaxy reads 'mostly harmless'. And in Kurt Vonnegut's novel The Sirens of Titan the entire history of Earth has been created as a spectacle to keep an alien messenger amused while he waits for a spare part for his spaceship. In fact, the joke is as much on him as us. The message he has been asked to deliver from one side of the cosmos to the other simply reads, 'Greetings'. Vonnegut, writing in 1959, also anticipates the Adams joke of a radically edited version of human experience. Owing to lack of archive and museum space, everything relating to the period between the death of Christ and 1 million AD is burned. A later history provided a brief summary of this period.

'Following the death of Jesus Christ, there was a period of readjustment that lasted for approximately one million years.'

Humans? So what?

But, assuming, for the moment, that we are not so dull, there are two serious reasons why we might never meet aliens, even if they exist: too much time and too much space.

The time problem is slightly more speculative. From the point of view of an advanced alien civilization, nothing happened on Earth for about 4 billion years. Then, suddenly, 100 years ago, it sprang to life with the generation of the first artificial radio waves. These would be detectable evidence of life. But they can travel no faster than the speed of light, so the first wave front will now be 100 light years away, not very far in cosmic terms. In fact, detectable radio waves probably only started being emitted in the 1930s with high-powered broadcasting. If aliens lived forty light years away, we could, therefore, hope for a response by 2010. One irony of these dates is noted in the film *Contact* (Robert Zemeckis, 1997). An alien communication is prefaced by some early television pictures of Hitler speaking. They are returning them to us to indicate their presence.

'Twenty million people died defeating that son of a bitch . . .' Cries the character Rachel Constantine, 'and he's our first ambassador to outer space!'

Meanwhile, as these waves were creeping outward at the speed of light, back on Earth there had been a staggering technological acceleration. Having developed radio, we went on to develop nuclear and biological weapons and to despoil the environment on an unprecedented scale. We also began to conduct experiments in high-energy physics that some say could cause 'a phase transition in the cosmic vacuum energy', which means they would terminate our existence and possibly that of the entire cosmos. We have, in short, rapidly acquired the means to destroy ourselves. In this context, intelligence, far from being the best gambit in the game of life, is, in fact, one of the worst. Being self-aware may lead inevitably to self-destruction.

This is accounted for in the Drake equation by the variable L – the length of time a communicating civilization survives. If, for example, the Cuban missile crisis in 1962 had led to a nuclear conflagration, then L would have equalled less than 100 years and the chances of two species contacting each other would be close to zero.

'Civilizations,' says a character in Stanislaw Lem's novel *Fiasco*, 'are harder to catch than a mayfly that lives for one day.'

In fact, Lem's point was not that civilizations would destroy themselves too quickly. Rather, he sees the idea of contact as the last pre-Copernican, anthropocentric delusion. It is based on the assumption that alien species would be like us to the extent that they would wish to seek out other species. But why should that be the case? Lem's point is that, beyond a certain level of maturity, civilizations simply lose interest in the search for others. So, even if we have cosmic contemporaries, it is absurd to believe they will wish to make contact. Having emerged from silence, civilizations return to silence. Again the window of contact is absurdly small.

Time, therefore, is as big an obstacle to contact as space. The sea of time is an immense expanse of, we believe, about 15 billion years. That, our physicists tell us, is how long it has been since the Big Bang created the universe. Of course, our physics may be wrong. But the likelihood is, if they are mistaken, they would be underestimating the amount of time there had been. We have become accustomed to the idea that aliens might exist elsewhere, the real problem might be that they exist elsewhen.

But it is space that is the more familiar problem. This did not, in the first optimistic phase of the Enlightenment, seem to be a problem at all. Certainly space was bigger than we had previously thought. The displacement of the Earth from the centre had created the possibility, embraced by Descartes, that space was, in fact, infinite. Nevertheless, to the later, Enlightenment mind, perhaps irrationally, it remained conquerable. This idea was given encouragement in the eighteenth century by the great ocean voyages of discovery. Captain James Cook encountered plenty of terrestrial aliens on his voyages and, indeed, was killed by some of them in Hawaii on 14 February 1779; he appears to have underestimated the growing tension between the islanders and the explorers.

Cook was the prime creator of the idea that we could travel outwards and find strange things and people; indeed, that it was our obligation and destiny to do so. In this he shored up the Enlightenment view that the world was ultimately knowable, that all things could be achieved by the ingenuity of man. The motto on his coat of arms may express the sceptical realism of the sailor – 'He left nothing unattempted' – but his place in the Western imagination is that of the hero who achieves whatever he attempts. Cook was also stylistically responsible for *Star Trek*. The British naval protocol used in the show arises from the sense that these wooden sailing ships – boldly going where no one had been before – were the prototypical exploratory vessels.

The rationality of the project of discovery, however, soon began to buckle under the strain of what was discovered. There was an ambiguity about the act of discovery. The strange people with their strange religions could certainly be seen as confirming the superiority of the Western, Christian model. But they could also be seen as demeaning its universalist claims. Perhaps we were destined to bring Christ to the natives, but how could we be sure? It might equally be the case that this perverse variety of beliefs and cultures was warning us of the arbitrariness of our own systems and convictions.

Romanticism turned its back on the rationality of the Enlightenment and, in doing so, transformed the voyage into something darker, stranger and more ambiguous. Coleridge's *Rime of the Ancient Mariner* is about a voyage to desolation and death. And, Edgar Allan Poe's *The Narrative of Arthur Gordon Pym of Nantucket*, published in 1838, was a tale of a terrifying and utterly inscrutable alien encounter. *Pym* is a fictional account of a voyage of discovery, but it was based on what was at the time a fairly respectable scientific theory called

Symmes' Hole. In 1818 John Cleves Symmes had issued a manifesto declaring that the Earth was open at the poles, and these openings led to habitable spaces within. (This idea remains potent. Many believe aliens reside beneath the ground or the sea, some say they have adopted these habitats as part of a secret deal with the US government.)

Pym's voyage leads him to the North Pole where he confronts the great chasm down which the ocean flows into the caverns beneath the surface of the Earth. There he sees an alien.

'But there arose in our pathway a shrouded human figure, very far larger in its proportions than any dweller among men. And the hue of the skin of the figure was of the perfect whiteness of the snow.'

The Other is a white nothingness to the human observer and the caverns beneath the Earth induce a very romantic vertigo, a sense of the abyss and of the narrowness of our quotidian vision.

And so, as the progressive vision of the Enlightenment faded, the possibility emerged that exploration was either a hopeless attempt to escape the boundaries of the self, or, worse, a form of impiety in the face of nature. At the beginning of M. P. Shiel's novel *The Purple Cloud* (1901) a preacher warns against further attempts to reach the North Pole.

'That there was some sort of Fate, or Doom, connected with the Pole in reference to the human race; that man's continued failure, in spite of continual effort, to attain proved this; and that this failure constituted a lesson – and a warning – which the race disregarded at its peril.'

The preacher turns out to be right. The hero reaches the Pole, where he sees a Poe-like vision of a circular lake with a pillar of ice at its centre – 'I had the impression, or dream, or fantasy, that there is a name inscribed round in the ice of the pillar in characters that could never be read; and under the name a lengthy date . . .' On returning, he finds mankind is being wiped out by a mysterious, poisonous purple cloud. As, apparently, the sole survivor he takes to burning cities in a fit of misanthropic rage. And, when he finally meets a woman, the only other person left alive, he declines to accept the role of a new Adam and to repopulate the planet.

'The modern Adam is some six hundred thousand years wiser than the first – you see? Less instinctive, more rational.'

Enlightenment humanism had given way to the misanthropy of modernity. Finally, in *The War of the Worlds* H. G. Wells completely reverses the voyage of discovery. Martian colonists arrive here and treat us with murderous disdain – their minds 'that are to our minds as ours to those beasts that perish . . .'.

But, ironically, soon after that book was written, the threat of alien invasion or, indeed, the possibility of any kind of contact, receded dramatically. The problem of distance suddenly seemed to become utterly insoluble. Einstein's theory of relativity was dependent on one absolute – the speed of light. Nothing could travel faster than about 300,000 kilometres per second. Quick as this is, it would still mean the nearest star was four years away and almost everything else hundreds, thousands or even billions of years away. Worse still, we couldn't even accelerate a ship to anything like light speed as the energy required would be enormous and, in any case, relativity also says that, at that speed, the ship's mass would have become infinite. The voyage of discovery that would simply incrementally annex ever larger regions of space was an impossible fantasy. The entire universe – except for the just about attainable speck of space in which we were imprisoned by physics – was beyond us. Impiety wasn't an issue any more. We couldn't even blaspheme effectively.

Einstein had dethroned both Newton and Captain Cook. But he had also raised the alarmingly frustrating possibility of a universe brimming with intelligent life forms that could never know of each other's existence. The speed of light was a universal limit. If we couldn't get to them, they couldn't get to us either. We would live and die in ignorance of each other.

Or perhaps we were being misled by the size of the universe. Of course it was big, but maybe it had to be that big. It took us 15 billion years to get here and that could be exactly how long it takes to create a single intelligent species. The universe is the way it is because we are observing; if it were otherwise, we would not be here, observing. This is known as the anthropic principle and, perversely, it restores man to his pre-Copernican position at the centre of the cosmos. We are, indeed, the point of the whole exercise, not because God says we are, but simply because we are here. In this case, space may be empty but it will have a purpose – the creation of us. If there were aliens, they would probably also be alone, but in a different universe.

An exotic form of hope had, however, appeared on the horizon. This was quantum theory. Relativity was about the behaviour of very large bodies; quantum theory concerned the smallest things that could exist. It exposed the fine structure of matter as a foaming chaos, beyond the reach of our common sense. Anybody not shocked by quantum theory, said Niels Bohr, hasn't understood it. Cause and effect seemed to vanish at the quantum level, time behaved strangely and entities could be both particles and waves at the same time. It seemed to be as true as relativity, as necessary to the continued existence of the universe, but it was in complete contradiction. The two theories remain incompatible to this day.

The reason quantum theory offered hope was its weirdness. If the universe really was this strange, then perhaps the speed of light wasn't an absolute limitation after all. There were strange and exciting consequences of the theory. Non-locality, for example, seemed to indicate that a change made to a particle at one position could instantaneously affect the condition of a particle at a distant location. A signal appeared to have been sent at greater than light speed. Einstein never believed this theoretical possibility, though, after his death, it was experimentally established. But, as Paul Davies points out, this seems to be a case of confusing communication with correlation. No signal is involved.

Meanwhile, other weirdnesses were also beginning to emerge from relativity. Great speed changed the passage of time. An astronaut flying for five years at close to the speed of light would return to find hundreds of years had passed on Earth. The theory also predicted the existence of black holes, gravity sinks from which nothing, not even light, could escape. Relativity also portrayed space and time as a continuum, a kind of cosmic sheet, which, presumably, could be folded or warped back on itself, making travel from one end of the universe possible if only we could, somehow, leap across the fold. Worm-holes, space-time anomalies predicted in theory, could offer the possibility of such a leap that would violate the spirit but not necessarily the letter of the Einsteinian law.

All that we needed, we now say, was a sure sign from the physicists that the absolute of light speed could somehow be relativized by some such device. Kurt Vonnegut parodied the idea by inventing 'chrono-synclastic infundibula', space-time glitches. Unfortunately, these seemed to result in the total disappearance of anybody who encountered them. They were there to say to humanity, 'What makes you think you're going anywhere?'

There was one other possibility, which was that Einstein was not so much wrong – we are not yet ready to consider that possibility – as partial. His theory applied to only half the universe. There was another half on the far side of the speed of light. The idea is captured in the film *K-Pax* (Iain Softley, 2001) in which a man, Prot, claims to be from a very distant planet.

'What if I were to tell you,' a scientist asks him, 'that according to a man who lived on our planet, named Einstein, nothing can travel faster than the speed of light?'

'I would say,' Prot replies, 'that you misread Einstein . . . what Einstein actually said was that nothing can accelerate to the speed of light because its mass would become infinite. Einstein said nothing about entities already travelling at the speed of light or faster.'

This is a reference to the theory of tachyons. These would be particles that existed through the looking glass of light speed. They would travel at many times light speed. Their optimum speed would be 100,000 times light speed, enabling them to cross the universe in less than 100 years. This appears to contradict Einstein, but, in fact, tachyons can be predicted from his special theory of relativity. Perversely, these particles move quicker as energy is removed from them and slower as it is added. The rather troubling feature of tachyons is that they could, in theory, reach infinite velocity in a finite time, at which point they would be everywhere at once. Tachyon fields, needless to say, appear frequently in *Star Trek* and Douglas Adams employs the possibility of being everywhere at once in his infinite improbability drive spaceship.

After Einstein and once it became clear that our solar system was uninhabited, aliens in both fact and fiction had to find ways to overcome the distance problem. In fiction the solution is usually just taken for granted, as Kingsley Amis observed.

'But most commonly, the author will fabricate a way of getting round Einstein, or even of sailing straight through him: a device known typically as the space-warp or the hyper-drive will make its appearance, though without any more ceremony than "he applied the space-warp" or "he threw the ship into hyper-drive".'

This is what happens in *Star Wars* (George Lucas, 1977), in which ships simply leap into hyperspace. There are also various ufological theories about how the flying saucers may have got here by folding space-time or by deploying various forms of gravitational energy. Rather more poetically, vast distances are spanned in David Lindsay's novel *A Voyage to Arcturus* (1920) by back-rays. This is a form of light which flies back to its source. Back-rays from one planet are transported to another in bottles and, when released, they can pull spacecraft back to the point of origin.

Communication is also a problem, of course, as electro-magnetic waves of all types are also restricted to light speed. In *Star Trek* the solution is 'sub-space' transmissions and in Ursula LeGuin's *The Left Hand of Darkness* there is the quaint-sounding Ansible, which provides simultaneous communication across intergalactic distances.

We have, of course, neither Ansible nor warp drive. Never-theless, we do try to make contact – either by detecting alien signals or by sending our own. Sending signals is intrinsically problematic since we cannot be sure they would be understood in terms of their content or, indeed, recognized as signals at all. All signals would, in effect, be coded since alien species cannot be expected to read any of our languages. But no code can contain its own decoding mechanism – if it did, the aliens would have to decode the code in order to decode the code.

This is an insuperable problem, though we find this hard to accept. In the film *Independence Day*, for example, the most improbable invention is not the vast alien craft, the demonic aliens nor their fearsome weapon, it is the idea that a human laptop can be used to upload a virus into an alien computer. Since we live in a world in which Apples can barely talk to PCs, this is madness. But, of course, it is based on the quasi-anthropocentric assumption that all computers are the same and that there can, therefore, be no insuperable coding problem.

And, in the real world, we continue to signal to the aliens in the hope they will understand. In the 1820s one Carl Friedrich Gass devised a method of signalling our presence that involved planting groves of pine trees in huge triangles. These would, he hoped, be seen by aliens as evidence of an ordering intelligence on Earth. The triangles, Gass argued, should be right-angled to indicate our awareness of Pythagorean geometry. The big triangles may seem culturally neutral, but there is a small chance that trees naturally grow in regular shapes on the aliens' planets. They would not, therefore, be a sign of intelligently guided planting. And, of course, there is a much greater chance that the aliens see nothing special about right-angled triangles.

In 1840 Joseph von Littow of the Vienna Observatory suggested digging a two-mile-long ditch in the Sahara, filling it with kerosene and setting it alight to form a beacon, visible from space. The straightness of the line would be intended to suggest this was not a natural occurrence – but even this may be subject to a decoding problem. Why should straight lines necessarily signal design to an alien mind? Fires, of course, often occur naturally.

In 1941 Sir James Jeans suggested using a searchlight to signal prime numbers to the Martians. Again there is an obvious decoding problem – also, of course, there are no Martians – but the assumption from then onwards was the language of mathematics and physics was at least likely to be universal. It is hard, the argument goes, to imagine an alien civilization advanced enough to pick up our messages which did not share at least some of our fundamental mathematical assumptions.

The belief that science and mathematics must form some sort of universal language is central to all attempts to either detect or signal to aliens. This is often extended to a conviction that even alien technology must be similar. A NASA report of 1971, for example, proposed Project Cyclops, a 1000-antennae system for detecting alien signals:

Regardless of the morphology of other intelligent beings, their microscopes, telescopes, communication systems, and power plants must have been at some time in their history, almost indistinguishable in working principles from ours. To be sure there will be differences in the order of invention and application of techniques and machines, but technological systems are shaped more by the laws of optics, thermodynamics, electromagnetics or atomic reactions on which they are based, than by the nature of beings that design them.

The idea is, of course, fantastically improbable. Do we know everything about the law of optics and is it certain that they can only result in one type of telescope? No and no. But the report does demonstrate how far technocratic conviction can go in anthropomorphizing the alien. Indeed, Project Cyclops itself had a quaint, hubristic madness about it. The 1000 antennae would be joined by a network of tunnels and would be run by staff who would live with their families in a specially designed town called Cyclopolis. The heroically collectivist imagery and the strangely Art Deco name confirm the view of the philosopher David Lamb that the whole SETI project differs 'from much of scientific activity in the sense that it is driven by metaphysical beliefs and deep psychological desires for companionship'. Cyclops, though a receiving station, could also have been a signalling beacon.

Another philosopher, Nicholas Rescher, has effectively taken apart these anthropocentric assumptions simply by examining Earth history. Human ideas and purposes change radically and frequently. Never mind trying to signal to aliens about, for example, our knowledge of quantum theory, we couldn't even signal to our own great great grandparents in those terms since they would have no idea what quantum theory was.

'Given intelligent beings,' writes Rescher, 'with a physical and cognitive nature profoundly different from ours, one simply cannot assert with confidence what the natural science of such creatures would be like.'

The even harsher judgment is that the very idea that any form of intelligence exists elsewhere is human wishful thinking – not just vainly hoping that we may not be alone but deeply in denial about our own contingency. Evolutionists tend to damp the fires of our extraterrestrial enthusiasm simply by pointing out that the evidence all points to how capriciously we stumbled into being. Evolution, said Stephen Jay Gould, is 'a staggeringly improbable series of events, sensible enough in retrospect and subject to rigorous explanation, but utterly unpredictable and quite unrepeatable'. In the early sixties, just as the US was about to spend large sums of money on exobiology, now known as astrobiology, the biologist George Gaylord Simpson produced a devastating essay called sombrely, 'The Nonprevalence of Humanoids'.

'The assumption that once life gets started anywhere, humanoids will eventually and inevitably appear is plainly false. The chance of duplicating man on any other planet is the same as the chance that the planet and its organisms have a historical identity in all essentials with that of the Earth through some billions of years. Let us grant the unsubstantiated claim of millions or billions of possible planetary abodes of life; the chances of such historical duplication are vanishingly small.'

Still, if search and signal we must, some assumptions have to be made. The central one, underlying all others, is that we do, in fact, inhabit the same universe. This is not as obvious as it sounds. An utterly different brain and perceptual apparatus may effectively mean an utterly different universe. But if, indeed, we do live in the same place, then there is a chance that we can meet in the same virtual space. This is known as the 'cosmic water hole' as it would simply be the place where different creatures would gather. It is, in fact, a radio frequency range – between 1400 and 1800 megahertz, a wavelength of 18–21 cm. This is naturally the least noisy part of the radio spectrum and it includes the 'Song of Hydrogen'. The most common element in the universe radiates at 21 cm. On the face of it, this is the likeliest place in which one species would signal to another. Of course, this too presupposes quite a high level of similarity between species. But we cannot look everywhere at once.

The hope is that civilizations become more detectable as they become more advanced primarily because they use more energy.

Nikolai Kardaschev, a Russian astronomer who led the first Soviet search for an alien signal in 1963, postulated the idea of super-civilizations, billions of years ahead of us. We simply use the power from the sun that happens to hit Earth; they might use all their star's energy output. Or, perhaps, they could exploit their entire galaxy for energy. Kardaschev classified our civilization, based on energy use, as Type 1; Type 2 could engineer an entire solar system; Type 3 could engineer galaxies. We could conceivably detect such civilizations by their anomalous energy patterns. The American scientist Freeman Dyson suggested a specific way of harnessing stellar energy – by constructing a huge sphere around the local sun. Attempts have actually been made to detect these 'Dyson spheres', though without success.

There is one further development that could have happened in advanced civilizations that would, in fact, answer Enrico Fermi's question – Where are they? The answer would be that they are all around us. Travelling at the speed of light is not difficult. It happens all the time. Light and radio waves travel at no other speed. What makes it difficult is accelerating mass to that speed. But information naturally moves at that speed. Television signals travel from Earth to a satellite and back again without any appreciable delay. In theory, everything is information, ourselves included. If we could, therefore, either download ourselves on to a computer or create a conscious computer program, then we could transmit this at light speed across the cosmos. This information could include the instructions on how to build a computer on which the program would run. If this is possible, then the overwhelming incentive for an advanced civilization would be to abandon any attempt to transport mass and to send computer simulations of themselves instead. These may well be whizzing past us now but we would have no way of recognizing them.

So far, therefore, there is nothing and continued lack of success in all these searches eventually had its predictable political effect. In 1994 the US government terminated funding for SETI.

'The Great Martian Chase,' said Senator Richard Bryan of Nevada in opposition to the spending, 'may finally come to an end. As of today, millions have been spent and we have yet to bag a single little green fellow. Not a single Martian has said take me to your leader, and not a single flying saucer has applied for FAA approval.'

Nevertheless, we continue to sweep the heavens and to send our signals. The modern phase of signalling really began on 16 November 1974 when Frank Drake sent a message from the mighty Arecibo radio telescope which described where we were and what we were like. It was sent in the direction of M-13, a promising star cluster in the constellation of Hercules. The British astronomer royal at the time, Sir Martin Ryle, surprisingly rebuked Drake on the basis that it might attract the attention of hostile aliens. Various probes that will leave the solar system carry hopeful messages in the faint hope that they will be scooped up by some intergalactic cruiser. Meanwhile, in the private sector, SETI continues its search. Now, through SETI at Home, it uses the PCs of volunteers to comb through the data retrieved through its radio telescopes. Each volunteer dreams of being the one to find the signal.

It may be, as some have suggested, that searching for an alien signal is like a drunk at night looking for his keys under a streetlamp, simply because that is the only place he can see. But the drunk needs his keys; do we need that signal in the same way? The answer appears to be yes.

The discovery of the immensities of time and space in which, dust motes, we hang has appalled and provoked our species. The discovery of the preposterous chain of chance that brought us here has made things worse. The scale of time and space asks us how we can possibly be at home and contingency tells us that we can't. But one inexplicable regularity in the electromagnetic spectrum that told us we were not alone would, at least for a moment, make us feel we were not quite so lost in time and space and not quite such a freakish arrangement of matter. The next moment, of course, we would discover new anxieties. At that point, paragraph 8 of the SETI Declaration of Principles Concerning Activities Following the Detection of Extraterrestrial Intelligence perhaps wisely demands a pause for thought.

'No response to a signal or other evidence of extraterrestrial intelligence should be sent until appropriate international consultations have taken place. The procedures for such consultations will be the subject of a separate agreement, declaration or arrangement.'

We couldn't get our leaders together to meet Klaatu. Imagine trying to hammer out our response to the aliens' 'Greetings!'

15

Marvin Minsky/Steve Jobs

The meat machine is what lies between your ears. Marvin Minsky, artificial intelligence researcher, coined this term for the human brain to make a point. His point was that, whatever else we may want it to be or hope that it might be, the brain is always and only a machine made of meat. The mind is just what happens when the meat machine gets going.

This has certain implications. If the brain is *no more* than a meat machine, then we should be able to replicate it, or something like it. There is no special mind stuff that lies beyond our manufacturing competence. And, since the mind is just what the brain does, as the kidneys secrete urine or the heart pumps blood, then that too should be replicable. Mind is just electrical activity in a few pounds of meat. Biology can have no monopoly on this phenomenon.

Of course, the brain may appear to be too complicated to replicate. We often say proudly that it is the most complicated thing in the universe. But we have no grounds for such a claim, partly because we don't know what else there is in the universe and partly because the brain may, in fact, be very simple, just a meat mechanism that follows a few simple rules. Following simple rules through billions of iterations can, as chaos theory has shown, produce strange and complex patterns. The outcome – the mind – may be complex, but the cause – the meat – may not. But anyway, even if it is genuinely complicated, it remains a finite lump of meat which can, in finite time, be replicated.

To think of the mind as the brain and the brain as meat is either glaringly obvious or wildly counter-intuitive or both. For most of human history it has been wildly counter-intuitive. Human beings have tended towards dualism – the division between matter and mind or spirit – because that is what it feels like to be self-aware. There is a world of matter to which we apply our minds. If they were the same thing, that sentence would make no sense and they certainly do not feel like the same thing. If they did, it is not clear we could think at all. Most animals are, perhaps, in this condition, as are human babies before they become aware they are but one thing in a plural world. Fully conscious adults, however, are obliged to move though the world in a condition of partial alienation, knowing they are both of it and yet not of it. To be self-aware is to be an alien.

Simply saying, as AI researchers and philosophers tend to do, that the mind is *just* what the brain does is not an argument against this dualism. If it is *just* what our brains do, why is it not *just* what the brain of a lizard does? Reptilian dualism has not, so far as I am aware, been demonstrated. We, plainly, experience something rather different, something that, apart from anything else, makes us reflect on that experience. Our innate dualism may, it is true, simply be the result of increased iterations of the patterns in the meat. It may also have been selected by evolution as an adaptive advantage. But that is simply to offer theories about how it happens. It is not to say what it is.

In fact, we have absolutely no idea what it is. In spite of the billions poured into AI research, we have come nowhere near producing anything remotely like a thinking machine. The most advanced computers remain inert, programmable lumps and the most ingenious robots can just about build cars or act vaguely like dogs. This, in spite of the fact that, for more than fifty years, scientists have been assuring us that an intelligent machine is just round the corner. Why they have so far failed is the subject of many other books that have already been written and some that have not. But that they have failed is indisputable.

Yet we still tend to think that the first fully accredited contact with an alien intelligence is likely to be between a human and a humanly constructed machine. This is understandable. Since World War II, machines – or rather their makers – have been engaged in a concerted effort to meet us and our desires on more or less equal terms. Amidst the anxious, consumerist euphoria of the fifties this drive manifested itself in labour-saving gadgets: washing machines, vacuum cleaners and so on. The machine had been moving into the house for some time, but now it was becoming thoroughly domesticated. These were, however, though clad in chrome and white, old-fashioned machines, carrying out familiar industrial processes in the home. The computer, for the time being, was hidden away in mysterious places. One estimate at the time was that no nation would need more than four or five computers.

All this began to change in the sixties when it became clear that advances in computing technology meant that computers could be made easily available and that the technology could be used more widely – in, for example, washing machines and vacuum cleaners. It took, however, another twenty years for the significance of this to sink in.

Perhaps the key figure was Steve Jobs and the key date was 24 January 1984, the day Jobs' company, Apple, launched its Macintosh computer with an advertisement in the TV coverage of the Super Bowl. The Mac was explicitly a product of the hippie era. It was produced by corporate dissidents who wanted to bring computing to the masses. Jobs was and remains a maverick in corporate America, a man whose explicit task has been to bring humane values and high aesthetics to his products in contrast to such seemingly faceless monoliths as IBM and Microsoft. This is idealistic in principle, but, in practice, it becomes a huge step towards the admission of the machine into our inner lives. It is, in fact, another cover-up.

The whole point of the Mac was the concealment of its machinehood beneath a friendly mask. The mask was the screen domesticated into a 'desktop'. A 'mouse' moved an arrow around this screen. The user pointed this arrow at an object and clicked on the mouse to make something happen. The complex of mouse–arrow–click was a kind of theatrical prosthetic designed to convince you that this was just like using your hand. This did not require you, as all previous machines had done, to talk machine language, it required the machine to pretend to talk yours. As social scientist Sherry Turkle has pointed out, the Mac emphasized surface manipulation combined with ignorance of any underlying mechanism so that people would no longer command the machine, but would rather enter into a conversation with it. Pointedly, the Mac device of the desktop and the virtual hand, now adopted by all computer operating systems, is known as a graphical user inter*face*. The machine has acquired a face to return our gaze. This was what we had expected to find only among the stars.

The whole mouse-click business has now become so familiar that it is easy to miss what a radical step had been taken. Talking to a machine while not knowing what is going on behind the screen is, in a limited but important sense, like talking to a person. We don't know what a person is thinking any more than we expect to know what an Apple Mac is actually doing. Concealment – a more negative but also more apposite word than interiority – is a first step towards personhood. And this is not just, as is often said, a virtual personhood, it is actual. The computer is actually concealing *itself*.

All computers now work in much the same way and, as far as possible, all other machines are converging on this conversational model. We are now accustomed to entering into conversations with cash machines, digital cameras, mobile phones and telephone answering systems. Even washing machines have 'displays' which are, in effect, alien faces that look back at us expectantly. None of this can be described as intelligent behaviour, of course, but it is a more effective *simulation* of intelligence than I think we have yet realized and it may, in the long run, be more conceptually significant than much current AI research. This is because the consumer machines are tuning their behaviour to ours rather than embodying an abstract theory of what we are like. They are subject to the evolutionary pressures of the marketplace.

The effect has been to immerse us in a ghostly but nonetheless real climate of otherness. A cash machine is possessed of a very deep interiority. It is the face of a vast, global computer system that can give or deny us money anywhere in the world. In crisis, we can appeal to a human being, but the system doesn't want us to because, if we kept doing that, the machine would have no purpose. We are aware of such systems as – for want of a better word – thinking about us.

Yet the use of the word 'thinking' remains, strictly speaking, metaphorical. It refers to the effect of machine behaviour on us rather than a property of the machine. In spite of the best efforts of the AI industry, computers still won't think for themselves. Plainly, the human brain – the only fully functioning thinking machine we really know – is nothing like a computer. The famous Turing test – in which if we can talk to a machine and not know that it was a machine, then we must credit it with intelligence – has not been passed. Maybe the reason is the 'Chinese room' argument of the philosopher John Searle. This, in essence, makes a distinction between appearing to think by following rules and actually thinking by applying an autonomous interiority to input material.

But these issues have become irredeemably stale for the simple reason that machines *as they already are* have bypassed them all. Both the Turing test and the Chinese room argument are highly anthropocentric. They overrate human consciousness. It is perfectly possible, for example, to imagine machines devoid of human-like self-awareness and utterly incapable of passing the Turing test which could, nevertheless, decide to overthrow a government. The mistake is to think the verb 'decide' requires human-like interiority. Animals make decisions all the time, as, in primitive ways, do computers. This laptop, periodically, 'decides' to tell me things I don't want to know. The human brain is nothing like a computer but why should the computer *need* to be like a human brain in order to become convincingly alien?

All of which is to say that the contemporary mind is, because of the mechanization of its daily life, already, though inexplicitly and unconsciously, in contact with aliens. The nightmare SF scenario – as in the *Terminator* films – is that these aliens will one day cut loose from human control and take over. At that point defeating the machine will become the only human imperative. The slightly less common but much more optimistic SF scenario is that it is the machines that will ensure the survival of consciousness into the future. The climax of *A.I.* (Steven Spielberg, 2001) indicates that this is precisely what happens – the machines outlive the humans and inherit the Earth. Perhaps we should not regret this too much; they are, after all, our machines, our offspring and, therefore, they are possessed of something like our consciousness.

The machine-takes-over story should not, in our present condition, seem that far-fetched. To the extent that it feels as though it is already happening, it *is* already happening. We already talk to machines, we already interact with – without being able to control – the non-human. One of the key reasons the alien is such a potent figure in the contemporary imagination is that we know it is already here.

This also prepares us for contact with a genuinely alien – in the sense of extraterrestrial – machine. The idea of alien robots like Gort in *The Day the Earth Stood Still* is very familiar. One of the things we assume technologically more advanced aliens will have done is create machine intelligence. We assume this because the intelligent machine has for so long been a futuristic fantasy. It is also reasonable to expect that, given the distances involved and the problems we have encountered in space, the aliens may have decided that it is easier to send machines on long voyages.

On the other hand, it is also possible that the aliens are machines. Judging by the future forecast by Spielberg's *A.I.*, the replacement of organic intelligence by machine intelligence may well be a normal evolutionary process. It may not, in fact, be an especially radical step. We are already beginning to turn ourselves into cyborgs – part artificial, part organic entities – through the use of heart pacemakers, artificial limbs and even reading glasses. Increasing numbers of such interventions are certain to become available. Indeed, cyborging is now seen as a good, charitable cause. The actor Christopher Reeve, who became paralysed after a horse-riding accident, has become a high-profile advocate for the improvement and wider application of cyborg technology.

Indeed, he is celebrated because his own cyborg nature has overcome his mere human limitations.

It is perfectly possible to imagine at some point in the very near future a self-aware entity that is more machine than human. There will then cease to be a clear barrier between human and machine. Rather we will come to perceive the distinction in the way we now perceive gender differences – as a continuous scale rather than an either/or division. Again this tends to render stale the old AI debates by supplanting them with the reality of the world as it is now.

In Kurt Vonnegut's *The Sirens of Titan*, human settlers on Mars are formed into vast armies to launch an extremely ill-judged invasion of Earth. The soldiers are loaded on to rockets possessed of only one control – an on/off switch. 'On' sends the ship on its pre-programmed trajectory. 'Off' does nothing. It is merely installed 'at the insistence of Martian mental-health experts, who said that human beings were always happier with machinery if they thought they could turn it off'.

Not to be able to turn the machine off is to cede control, to accept its alien power. Welcome aboard.

16

Alien Metal

'. . . The pity of it smarts/,' wrote John Ashbery, 'Makes hot tears spurt: that the soul is not a soul,/ Has no secret, is small, and it fits/ Its hollow perfectly . . .'

Illness expresses itself in the symptoms of the age and ours is, among other things, the age of the disordered machine.

In 1959 in *Scientific American*, the psychiatrist Bruno Bettelheim recounted the case history of a schizophrenic boy named Joey. This boy had, in his imagination, turned himself into a machine that needed electrical power in order to function. Joey would turn himself 'off' – sitting idly in a corner – or 'on' – exploding into life and rising through higher and higher gears until he exploded with cries of 'Crash! Crash!'

He functioned as if by remote control, run by machines of his own powerfully creative fantasy. Not only did he himself believe that he was a machine but, more remarkably, he created this impression in others. Even while he performed actions that are intrinsically human, they never appeared to be other than machine-started and executed . . .

A human body that functions as if it were a machine and a machine that duplicates human functions are equally fascinating and frightening. Perhaps they are so uncanny because they remind us that the human body can operate without a human spirit, that a body can exist without soul.

In *Westworld* (Michael Crichton, 1973) robot cowboys provide holidaymakers with a simulated Wild West experience. The robots are programmed not to harm people. As a result, apparently hardened gunslingers are defeated in shootouts by urban hicks. The robots are there to satisfy the violent fantasies of the bored middle classes.

But then they begin to kill people. The whole system of the resort has gone wrong; the scientists say it is suffering from central mechanism psychosis. The robots cannot even be turned off. The purpose of the gunslinger – to act like a murderous outlaw – has not changed; the restraints on his behaviour – not to kill real people, to stop when instructed – have simply been removed.

The truly chilling aspect of the disordered machine that pursues the heroes of the film is its relentlessness. Not only can it not be turned off, it also cannot be reasoned with, an uncanny effect in view of its very human aspect. Furthermore, of course, it pursues its disordered course with error-free efficiency.

The robot cowboy also goes on functioning when most of its simulated human exterior has been stripped away by fire. Significantly, one part that isn't stripped away is the face. And so we have the uncanny image of a human face behind which is only machinery.

It is a recapitulation of the climactic scene in *Metropolis* (Fritz Lang, 1926). False Maria – a robot made to look like the beautiful woman who aspires to save the workers of the city – is burned upon a pyre. As the flames reach her, her flesh melts away to reveal the machinery beneath.

But the *Westworld* version with its focus on the face had a specifically postwar flavour. It was to be highly influential, the image of the machine behind the face being used repeatedly in later movies. Spielberg deploys it several times in *A.I.* The detached, machine face has a uniquely shocking power as it denies us the consoling depth we impute to real faces, the depth we often attribute to the presence of a soul. But also, less night-marishly, it relates directly to the GUI – graphical user interface – developed for computers in that, like the virtual desktop, it is a familiar, acceptable human feature used to conceal the workings of a machine. Conceptually, it is the face on the machinery in *Westworld* that might be said to anticipate the Apple Mac.

What is uncanny about the image is the shocking sense of disconnection, the disjunction between flesh and metal. Steve Wright, an American comedian, used to tell a one-line joke – 'I once saw a man with two artificial legs . . . And real feet.' The real feet are uncanny because disconnected from the real body by metal. We feel that they cannot be so disconnected and remain 'real'. The joke depends on the pauses just before and just after the phrase 'and real feet'. The pauses are a simulation of the disconnection. The disconnection felt at the sight of the face mask covering the machinery arises from the fact that faces are what we use most commonly and intimately to identify ourselves. Faces are always taken to express a mind, if not a soul, that lies behind. But, like the feet at the end of the artificial legs or the user-friendly screen of a computer, these faces are connected only to machinery.

Such examples point to the twin horrors of the disordered machine: absence and loss of control. They are, in fact one: the horror of the humanly created alien. To Bettelheim, his patient Joey suggests that there is nothing behind the façade of the body but machinery. Joey made the machine from which Joey then absented himself. There is an absence in the hollow that should contain the soul. The robot cowboy goes out of control and cannot be stopped. Behind his body there really is nothing but machinery. He is out of control because he has no soul, he cannot be reached, reasoned with. The empty, soulless machine is a horror because of what it might do to us. But a far worse horror is the one glimpsed by Bettelheim – that the machine, soulless and out of control, might be us.

Descartes thought that animals were machines. This was a specific version of the general scientist view that the world is a decipherable, mechanical system of cause and effect. If this is truly so, then free will and consciousness must be illusory, as must the conviction that there is any difference between living and non-living matter. Descartes agreed up to a point. The world was, indeed, a mechanical system, but humans, he said, were different. We had souls, reason, free will. He was, at one level, simply reacting to Montaigne, who had insisted that the beasts were our equals, if not our superiors. This idea has a distinct intellectual history from Montaigne, down to, in our time, John Gray. In the 1930s the anthropologist George Boas gave it a name – theriophily. Theriophilists variously argue that animals are as rational as humans, or less rational than humans but better off without reason, or, indeed, they are more rational than humans. Or they may say that animals are happier than humans, in that nature is kinder to them and does not so thoroughly thwart their desires. Or, finally, they may say that animals are more moral than humans; animals do not, after all, organize genocides. For a theriophilist, to say animals are machines and we aren't is absurd anthropocentrism.

But, for Descartes, animals just react; in contrast, we reason and make decisions. The distinction is decisive. Thanks to Copernicus and Galileo, human beings may no longer be special in the cosmos, but, thanks to Descartes, they could, at least, be special on Earth. Not for long. For the next 400 years, the corrosive acid of the modern science of which Descartes was the philosophical progenitor was to shrivel this last fragment of specialness almost to nothing. Now, it seems, we may well be machines. And, even if we aren't, we soon shall be.

The first act of modern science was the placing of a machine, a telescope, to the eye of a man, Galileo. He cyborged himself to become more than what was generally regarded as human at the time. He invented the experimental method, a method that, typically, requires an apparatus to be applied to the fabric of the material world.

Logically, this method – assuming the world is made only of matter – must ultimately be wholly successful. The world must, finally, be exposed as a mechanism. In that case, the being that perceives the world must also be a mechanism. Before Galileo, this was an idea that had been nurtured as a possibility in the anatomical drawings of the Renaissance. Leonardo da Vinci looked at the anatomy of arms and legs and saw they were like machines. He drew them as if they were, his flayed arms anticipate the robot arm ripped open to reveal the metal beneath in *Terminator 2*.

But the experimental method added a materialist theory to justify this similarity between flesh and machine. Two moments, separated by over three centuries, are critical. A few years after Galileo saw mountains on the moon, William Harvey discovered the circulation of the blood. The unravelling of the molecule of DNA by Watson and Crick in 1953 was similar in imaginative impact. Harvey had seemed to show the body was one kind of machine – a kind of hydraulic pumping system – Watson and Crick showed it was another – a computer-like information processor. Either way, the message was clear: the body is a machine and, certainly in the minds of the militant atheists Watson and Crick, there was only the body and, therefore, we were machines. We reacted like a machine, our thoughts were programmed and we lived and died as machines.

This slowly developing but irresistible idea of human machine-hood is a crucial aspect of our contemporary sense of the alien. Indeed, it might be said to be the idea that created the whole genre of science fiction, the imaginative habitat of the alien.

In his book *Trillion Year Spree* (1986), Brian Aldiss, author of the short story on which Spielberg's *A.I.* is based, recounts the history of SF. Previous histories had invariably traced SF narratives back to the ancient world, to, for example, the account of a voyage to the moon by Lucian of Samosata in the second century AD. The obvious reasoning was that what was distinctive about the genre was the element of the fantastic in accounts of voyages to other worlds or encounters with other beings. But this is not really adequate in that it leaves out the word 'science'. A better name for this writing would be fantastic fiction. For critic Adam Roberts, what is truly distinctive about SF is that it 'requires material, physical rationalization, rather than a super-natural or arbitrary one'. SF may invent its own science, but it remains, definably, science. Lucian never thought he could really go to the moon, but we have done so and we suspect we will soon know how to go further.

Similarly, Aldiss does not accept that ancient fantasies are truly SF. Instead, he persuasively insists that the roots of the genre are in the romantic era. It was a period when technology and new scientific knowledge were pressing in on the human imagination as never before and when the first attempts were made to turn it into art. Specifically, Aldiss sees Mary Shelley's *Frankenstein: or The Modern Prometheus* (1818) as the great SF precursor. Frankenstein is about the creation of one human being by another. If new technologies can do this, then, in fact, their creative powers are limitless, comparable to those of God. The monster is a poetic version of what Bettelheim saw in Joey. Being made by man, he cannot have a soul for that can only be the gift of God. Yet, like Joey, he plainly lives. And, worst of all, he dies because he is betrayed by his human maker. Not only does this creature lack a soul and yet live, it also exposes the iniquities of which beings with souls are capable. Why, in that case, bother with a soul? Why are we so special?

Aldiss shrewdly notes that *Frankenstein* was published at about the same time as Lyell's *Principles of Geology*, the book that announced we were lost in the abyss of deep time. The horror of the abyss and the horror of the soulless, machine body had converged to create SF.

Nevertheless, 200 years after Shelley and 400 years after Descartes, humans remain ambivalent about their machinehood. We seem to have both accepted and rejected the idea. We have accepted it in the sense that we assume that our bodies and our minds can be fixed like a machine. Spectacles and hearing aids may seem harmless enough, but they do embody the assumption that we can see and hear the world better using mechanical means. True enough, but what do we see when we remove the glasses? A different world, another world, our own world? And whose life is it exactly when a pacemaker has to keep it going? Many of us, most of us, live, these days, on borrowed time, time borrowed from technology.

Psychopharmacology – especially in the use of drugs like Prozac – takes this much further in that it changes the way we feel, making machines of our sensibilities. Since, as Ludwig Wittgenstein observed, the happy man sees a different world than the sad man, this is to change everything. Prozac makes new worlds. It also implies that a whole world view can be treated as a mechanical issue. Taking Prozac becomes like defragmenting a computer hard drive, simply a way of improving performance. Enthusiasts for Prozac have spoken of 'listening' to the drug. But, as Sherry Turkle has asked, to whom is one supposed to be listening? The answer is: to me, the alien machine that suffers no depression.

Explicitly, however, we tend to reject the view of ourselves as machines. It is, when baldly stated, insulting. We have, we feel, some quality best described as a soul and this quality is desirable. 'Soul' expresses what we take to be the intangibility, the unreachability, of our inner lives. It is what is most real, but also what is most indecipherable to others.

Defending the soul has been an issue ever since modern science began. This has not been easy. The progress of science has steadily disproved vitalism, the belief that there was some force within living matter that made it different from the rest of the stuff of the world. As a result, if there is a soul, biology has not found it in our bodies, just as astronomy and geology have failed to discover either a heaven above or a hell beneath. In spite of this, our dualist nature still demands that there must be something else, some immateriality that suffuses our bodies with life. The story of Frankenstein pivots on this issue.

But the scientist faith in absolute materiality cannot let matters rest there. If it is true that there is such a thing as an immaterial soul, then science, an account of the behaviour of matter, must fail in its attempt to become a complete account of the world. The distinctively human will remain beyond its grasp. This idea was repellent to the new materialist faith in the experimental method and so, having conquered matter, science applied itself to the conquest of mind. Through biology and AI, science attempted to take the battle to the last stronghold of the dualist imagination, the mind.

The history of science thus spirals inevitably towards the stuff of the brain. It was the last unconquered territory. Through Charcot and Freud down, in our time, to the artificial intelligence researchers and philosophers, science manoeuvred itself into

position for the final assault. Meanwhile, the computer, supposedly the human mind's mechanical analogue, came to embody
the idea of the mechanical brain, which, if it could be made to
work, would signal that the land was, indeed, conquerable.

Once again this is a process that seemed to accelerate in the
immediate aftermath of World War II. In 1948 Norbert Wiener
coined the term cybernetics and produced his manifesto
'Cybernetics: or Control and Communication in the Animal and
the Machine'. This was a history of automata divided into four
stages: the mythic, the age of clocks, the age of steam and the age
of communication and control. The shift from the third to the
fourth was marked by a move from energy to 'the accurate
reproduction of a signal'. The models of the human body implied
by these stages were: the magical clay figure, the clockwork
machine, the heat engine and the electronic signal.

Wiener believed that the analogies between machines and
living organisms had become so clear that it was idle to insist on
a distinction. He believed it was 'best to avoid all question-
begging epithets such as "life", "soul", "vitalism", and the like,
and say merely in connection with machines that there is no
reason why they may not resemble human beings in representing
pockets of decreasing entropy in a framework in which the large
entropy tends to increase'.

Wiener was ushering in the fourth age in which information
flow and control would be seen as the machine analogues for life.
These would be the ways in which matter would be organized in
a lifelike way. At their heart would be the computer.

He was, of course, anticipating by many decades the technology that would be able to achieve this. The only computers
around at the time were massive and crude and the naïve but
logical initial response to the idea of the machine brain was that
it would have to be very big to mimic the capacities of the human
brain. The early valve-driven electronic computers, constructed
in the forties, were vast, not smaller and not that much more
effective than the great brass difference engine, the mechanical
computer proposed by Charles Babbage in the 1830s. The
obvious assumption, in the circumstances, was that a mechanical
brain would have to be gigantic.

In Clifford D. Simak's 1949 story 'Limiting Factor', astronauts discover an Earth-sized, apparently artificial planet with a metal surface fourteen inches thick. On the surface is a deserted city, but, beneath, they find a mass of machinery. The planet is not artificial, it just appears so because its entire surface is covered by this machinery to a depth of twenty miles. The astronauts decide the abandoned machine must have been a calculator – 'They were trying to work out an answer to the Universe, what it is and where it might be going.' The machine would not only be big, it would also be capable of answering questions its maker could not, questions previously thought of as metaphysical. The aliens had failed because of the 'limiting factor'. The machine had to be so large to perform its calculations that they had filled the entire planet and their steel structure could not bear any more weight.

This notion was rapidly rendered quaint by the development of silicon chips to replace the banks of vacuum tubes the first computers had required. The chip shrank roomfuls of tubes to the size of a fingernail.

But, in fact, even before the advent of the chip, SF writers had generally worked on the assumption that a manageably sized mechanical brain must be possible. In most SF, robots, typically, have portable brains that can easily be removed. In *Saturn 3* (Stanley Donen, 1980), the 'demigod' robot called Hector has to have a tube of brain tissue inserted into his body before he can function. And 'Quick, get his brain out!' they cry when Hector turns rogue. There is a compromise here with dualism. Even in the robot, the brain stuff is still different from the body stuff and can easily be removed, rendering the body useless. The robot as intelligent being is different from the computer. The robot contains a computer 'brain' which is its intelligence. The computer is solely its intelligence. The robot remains a dualist; the computer attains monism. The price the computer pays is that it cannot necessarily act intelligently. It may beat you at chess, but it can do nothing to save itself if, as a result, you toss it into the river.

Computers, however, crash and robots, like Hector, go mad. In fact, SF writers and film-makers often cling on to this idea of robots going crazy as evidence that human beings are still different. This is irrational because human beings also go crazy, but the emphasis with robots is on the *way* they crack up. The behaviour of Amee, the dog-like robot in *Red Planet* (Antony Hoffman, 2000), is not guided by any overriding principle – such as, for example, the 'don't kill humans' command programmed into the Westworld gunslinger. Rather, Amee has different modes, each of which functions according to a different principle. When it is accidentally 'flipped', it goes into kill humans mode. The inflexibility of the machine mind is consoling evidence that it can never match our adaptive competence. On the other hand, this very inflexibility may mean the machines can destroy us.

But why should they do that? In Amee's case, the robot is simply 'flipped' so no decision is made. But at the heart of the three *Terminator* films – *The Terminator*, *Terminator 2: Judgment Day* (James Cameron, 1984 and 1991) and *Terminator 3: The Rise of the Machines* (Jonathan Mostow, 2003) – is the idea that the acquisition of consciousness by a machine at once fills it with the desire to destroy humanity. As the character Kyle Reese explains in *The Terminator*, 'Defense network computers. New . . . powerful . . . hooked into everything, trusted to run it all. They say it got smart, a new order of intelligence. Then it saw all people as a threat, not just the ones on the other side. Decided our fate in a microsecond: extermination.'

Skynet, the computer, becomes 'self-aware' and launches a nuclear attack designed to claim the world for the machines. Skynet's value to its human designers is precisely its autonomy. It can fight wars infallibly without human involvement. But self-awareness renders this autonomy catastrophic. Skynet seems to be a dispersed system, having no core – suggesting its self-awareness must also be dispersed – and, therefore, there is no plug to pull and no 'off' switch. Nevertheless, human beings are its potential destroyers and so it chooses to destroy them first. The idea is obviously derived from the internet. This, too, cannot be turned off, is coreless and dispersed. It is everywhere and nowhere.

Less pessimistically, the fallen machine can redeem itself by restoring its metal soul to the bosom of humanity. Hal, the computer that controls the spaceship in *2001: A Space Odyssey*

(Stanley Kubrick, 1968), goes mad and kills all but one of the crew. Its reasons are unclear. But in the sequel, *2010* (Peter Hyams, 1984), Hal is found to have been justified because he was driven mad by his human programmers. Also he redeems himself by sacrificing himself for humanity. His programs had inserted contradictory programs that required him to conceal the purpose of the mission as well as to both put the mission first and put the survival of the crew first. Hal, like Robby or Gort, is too pure. His machinehood lacks the murk and the compromise of our humanity.

The most consistent and certainly the most well-known expression of the machine-soul conflict is, inevitably, in *Star Trek*. In its *Next Generation* version, the crew of the *Enterprise* includes Commander Data, a robot. The fact that he has the rank of commander suggests he is both a very effective human simulacrum and that his intentions are entirely benign. On the whole, this is the case. He does, occasionally, go berserk, but not in ways that ultimately threaten his authority or compromise his essential goodness.

Simply as an idea – the idea being the possibility of logical purity – Data replaces Mr Spock from the first series of *Star Trek*. Spock, a Vulcan, was devoid of emotion and committed – to the horror of the emotional ship's doctor 'Bones' McCoy – to a strictly logical view of every problem. He was not actually a robot, but he represented a machine-like consciousness which judged sternly but usually kindly the emotional outbursts of his human comrades. In fact, perhaps because he was half-human, he didn't represent pure machine logic very well as he was plainly riddled with emotion, not least in his devotion to Captain Kirk, and Vulcans appear to have some very illogical, quasi-religious practices which periodically involved Spock putting on some suspiciously priest-like robes.

Data, however, is all robot and his precise status is constantly being debated. Usually, this involves an informal discussion about whether he can get jokes or understand human emotions. But, in one episode – 'The Measure of Man' – it is formally debated in a court of law to establish whether he can be accorded rights in the same way as an intelligent organic being.

Invariably Data wins this kind of debate, at least when it turns serious. In practical terms, he must win because he is a core member of the *Enterprise*'s crew and it would be hard to imagine a central character – and hero – of the series being so fundamentally downgraded. The overtones of racism would be intolerable. But he must also win because of the generally progressive, humanist mood of the series. Humanity must be generously extended to embrace this creature, just as various rights and protocols are accorded to the alien beings encountered by the *Enterprise*.

However, paradoxically, there is never any question that being human is a superior condition to being an android. Data constantly aspires to understand and to emulate humanity and the crew encourage him to do so. Frequently he is unable to cope. When his 'emotion chip' is inserted, human feelings overwhelm and paralyse him. And when the Borg Queen in the film *Star Trek: First Contact* (Jonathan Frakes, 1996) cyborgs him by connecting a piece of human skin to his arm, he is overcome by the erotic sensations he experiences, even though we are told, frequently, that he is sexually fully functional even without the skin. Plainly like a Stepford wife or the robot whores in *Westworld*, he usually functions without 'real' feeling.

When Data finally dies, in the film *Star Trek: Nemesis* (Stuart Baird, 2002), the tributes paid to him by Captain Picard are based on his heroic urge to better himself: ' "In his quest to be more like us, he helped us see what it is to be human." '

It could not have been said of a dead black crew member that he helped the rest of them see what it is to be white. In truth, like the droids R2D2 and C3PO in the *Star Wars* films, Data's status is closer to that of a pet. I suspect the underlying position is that, much as we may love Data and wish to accord him rights, he remains our creation. He is not, as we are, either self- or God-created, even though, at times, he might prove better than us.

This suggests a tension that cannot quite be resolved by the liberal imagination. We want the machines to be like us, but, on the other hand, we don't. It is present not just in fiction, but also in the minds of the technocrats.

'I wish,' said Rodney Brooks, the most celebrated AI

researcher at Massachusetts Institute of Technology, 'to build completely autonomous mobile agents that co-exist in the world with humans, and are seen by humans as intelligent beings in their own right. I will call such agents creatures.'

'Creatures' are created ones. We are God's creatures but we are not, like these robots, man's creatures. The distinction draws the necessary line in the sand. Dualism cannot quite be eradicated.

Thus Data the machine is kept in his place as an aspirant human. His ambitions draw attention not so much to his own good character as to that of the humans that made him and are now prepared to tolerate him as one of them. In spite of the competence, power and intelligence of the machine, the supreme mode of existence remains being human. And this supremacy is celebrated by our generosity in according quasi-humanity to the android. The soul survives intact, embodied only in human flesh.

In contrast, the soul does not survive in the Borg, the collectively minded enemy with which Picard is at constant war. The Borg are cyborgs in that they are both flesh and machine and they 'assimilate' other beings by implanting machinery into their biological bodies. The Borg are the reverse of Data. He wishes to move, at least spiritually, from machine to organic; they wish to push the organic in the direction of the machine. This is a threat to the assumed superiority of the organic and the fact that the Borg have a single, collective mind indicates that the true target of this threat is human individuality, the very quality Data so assiduously seeks to understand and embody.

To the extent, therefore, that the alien is a machine, *Star Trek* wishes to destroy – or, in the case of Data, convert – him in the name of humanism. The organic aliens in the series are accorded respect because they are organic, but they are all too plainly little more than realizations of desirable or undesirable human traits. The Klingon are Viking-like warriors, noble but dangerous, the Romulans fascists and so on. They are treated as bad humans, with diplomacy and, when necessary, lethal force.

The organic-better/machine-worse assumption is not, however, always so easily sustained. Agent Smith, the computer program in *The Matrix* (Andy and Larry Wachowski, 1999), finds us disgusting and both Gort, Klaatu's robot in *The Day the Earth Stood Still* (Robert Wise, 1951), and Robby in *Forbidden Planet* (Fred McLeod Wilcox, 1956) are, like Hal, more morally pure than humans. The problem is that the Picard position is based on a metaphysic of human goodness which is hard to sustain against the reality of human behaviour. Only by externalizing all human malice and projecting it on to the organic aliens that populate the *Star Trek* galaxy can Picard argue for our superiority to the machines, a superiority based on our spiritual balance between reason and emotion.

But, if the flaws remain in the humans, then we might reasonably find ourselves aspiring to the purity of the machine. After all, the flaws that most urgently need rectifying are not moral or spiritual but physical. Machines have easily replaced parts and, therefore, offer eternal youth, if not immortality. Female breasts, it seems, are always sexier when perkier, even when we know there is a silicon bag inside. (Cosmetic surgery may, in fact, be monistic science's most blatantly dualistic act. The very fact that it happens at all is an acknowledgement that mind and body are out of step. Older people are often said to possess the virtue of being 'young at heart', meaning their youthful inner self is not expressed by their decrepit outer self. Mechanically correcting the outer self to bring it more in line with the inner must mean there are at least two entities in conflict within the patient.)

Psychopharmacology extends this correcting of flaws to the realm of the mental, at which point it becomes impossible not to start asking an awkward question, the same question asked by Mary Shelley in showing her soulless monster being betrayed by ensouled humans. The question is: what's so special about being human, about *not* being a machine? The predictable modern answer – 'nothing' – has come in many forms, but all have been driven by the development of the computer.

Within AI research the technology of the computer has often produced a climate of wild irrationality. Most obviously, many have been driven by the conviction that the human brain must be an information processing system like a computer. But this is to assume that our computers are, somehow, inevitable and objective realizations of the only possible mathematics. It is this assumption that makes it possible for the laptop in *Independence Day* (Roland Emmerich, 1996) to talk to the alien mainframe. All computers *must* be the same, they can be built in only one way, *our* way. This was, in the early days of computing, reinforced, by behaviourism in psychology, the belief that discussion of all inner mental states was to be excluded as irrelevant. There was no internal state, there was only the observable nexus of stimulus and response. Thus, if we built a machine with sufficiently elaborate responses, it would be as 'conscious' as we are. 'Conscious' becomes, in behaviourism, merely a convenient term for a distinctively complex accumulation of responses to stimuli.

But, as the computer developed, the rhetoric of machine intelligence began to change. It became impossible to speak of these machines without also speaking of some kind of inner state. Most decisively and, now, most familiarly, we speak of computers as having 'memory'. For behaviourists, memory could only be understood in terms of the act of remembering. But memory in a computer plainly is an inner state, it would be deeply irrational to think otherwise. I can, after all, turn off this machine and then turn it on again. This document would reappear. Where has it been in the interim if not inside the machine? To argue that it has been summoned into existence just by the act of remembering is simply too bizarre to be worth considering.

If computer scientists had stuck to behaviourism and insisted memory was not an inner state, then I would probably be writing this on a typewriter. My computer's memory persists when I walk away from it, it even persists when I switch it off and it is certainly always there, retaining the details of what I am doing as well as of what I am not. It may be excessively anthropomorphic to call the inner state of computer storage 'memory', but the point is that somebody did so and it stuck. This will, I believe, turn out to be an act of naming comparable in significance to Adam's naming of the animals in Eden.

As a result of the development of the chip and of memory storage, the 1980s saw not just the advent of the 'user friendly' Apple Mac and the democratization of computing. The decade also saw the rise of video games. These had appeared in the 1970s and had inspired people to imagine the possibility that computers would no longer be mysterious corporate or governmental machines, but would enter everyday life. In the eighties the games became highly immersive experiences that represented a new level of the human–machine relationship.

In one sense, they may be said to have created a contest between human player and machine. But, in fact, as they grew more sophisticated, it became clear that the machine was taking the player to a third realm that was independent both of the player and of the machine itself. Obviously, the very fact that you could play many games on the same machine indicated that the computer wasn't your opponent. But, if that was the case, who was? The banal answer was that it was the logic of the game created by its programmers. But increasing interactivity meant that any such logic could not possibly be the sole creation of the programmers, it must emerge in the act of playing. So each playing of each game would have an independent logic of its own. It is no great leap of the imagination to change the word 'logic' in that sentence to 'life'. Indeed, it was a computer game – Steve Grand's Creatures – that seemed to confirm that some kind of autonomous, 'living' entity could be created inside a machine. Latterly, the game Sims built on this by creating a narrative environment that is, comically, both like and unlike the real world. The machines have created a Disneyesque super-reality with the added novelty that we can live there.

It is significant in this context that in *Star Trek: The Next Generation*, the *Enterprise* has a recreational place called the 'holodeck'. This is a place where computer programs can be run which totally immerse members of the crew in alternative realities. They can meet Leonardo da Vinci or they can be eighteenth-century sea captains. They can be Western gun-slingers, though, as in a normally functioning Westworld, they come to no harm. But, also as in *Westworld*, however, this can go wrong. The fact that the holodeck is such an easily assimilable concept indicates the extent to which computer games have penetrated the culture. By making machine life at least *seem* real, computer games weakened the old distinction between soulless machines and ensouled humanity.

In the nineties, the internet was to weaken this distinction further. The first great net joke was: 'On the internet nobody knows you're a dog.' When you play a computer game you can become somebody other than yourself in relation to the program – a creature within the game. But, on the internet, you can become somebody else *in relation to other people*. The net creates spaces – 'rooms', 'sites' – where people can gather and talk. They may even use visual identities – avatars. But they don't need to be themselves, they can be anybody. They can create a life they never had. Sherry Turkle writes:

> In the story of constructing identity in the culture of simulation, experiences on the Internet figure prominently, but these experiences can only be understood as part of a larger cultural context. That context is the story of the eroding boundaries between the real and the virtual, the animate and the inanimate, the unitary and the multiple self, which is occurring both in advanced scientific fields of research and in the patterns of everyday life. From scientists trying to create artificial life to children 'morphing' through a series of virtual personae, we shall see evidence of fundamental shifts in the way we create and experience human identity.

The most fundamental shift of all is to the acceptance of the creation and experience of identity within the machine. This has been anticipated by the Apple Mac and by its more successful imitator, Microsoft's Windows operating system. The point about using 'windows' on the screen is that they create virtual depth. Many windows can be open at once and the user can switch rapidly between them. The machine replicates the mind's behaviour, transferring attention rapidly from one thing to another. And, most importantly, in doing so it downgrades the sense of a unitary self. Instead, we feel there are multiple selves, each specially created for each different object of attention. The internet literalizes this further by actually placing the selves, not just the objects, on the screen. On the internet nobody knows you're a dog and even you can't be sure.

This is not yet the creation of machine intelligence, but it sets the scene in many different ways. For example, it seems to bring closer the day when we can download the content of our minds on to a computer. Already we can download bits of ourselves and the virtual self that is created can, in certain limited ways, learn more about us as we go along. Software can fill in forms for us because it 'knows' our personal details. The idea of a machine self that is our own or, indeed, wholly alien is thus brought closer by our imaginative acceptance of the possibility.

In the eighties the idea sprang to life. The increasing success and sophistication of video games and the appearance of personal computers were the two technical reasons. But there was a cultural reason – the exhaustion of modernism. The escape from the bourgeois and geopolitical oppression of the fifties into the carnival of the sixties had been a revitalization of modernism's utopian aspect, its continued belief in the progressive project of the Enlightenment. Hippies were the reincarnations of the *philosophes*, prophets of a world free of ancient restraints. The idea clung on through the seventies, but, in Britain at least, foundered in 1979 when the summer of love was replaced by the winter of discontent.

The eighties cult of the free market was a reaction to the failure of all these progressive utopian dreams. In the free market it was everyone for themselves and the greater good was to be served by the large-scale statistical effects of such behaviour. Individuals need not embrace any ideals, merely pursue their own best interests and all manner of things would be well.

The implication is clear: what is actually required of individuals is that they become machines. Hardline free market economics are plainly based on this assumption. Free the individual from all restraint and pretence and they will act as predictably as a machine. Indeed, so to act is the only ethical course since it is the only way in which society as a whole will be enriched. Anyone needing a private ethical or religious system could choose their own from the supermarket of postmodernity.

Even if you weren't a free marketer, this created a climate in which the idea of machinehood took on a new, seductive quality. The prime seducer was the novelist William Gibson. He first saw what was happening when he looked into one of the new video-game arcades.

> I could see in the physical intensity of their postures how *rapt* these kids were. It was like one of those closed systems out of a Pynchon novel: you had this feedback loop, with photons coming off the screen into the kids' eyes, the neurons moving through their bodies, electrons moving through the computer. And these kids clearly *believed* in the space these games projected.
>
> Everyone who works with computers seems to develop an intuitive faith that there's some kind of *actual* space behind the screen.

In 1983 Gibson published *Neuromancer*, the novel that was to define the new SF form of cyberpunk. In fact, reading *Neuromancer* remains almost the only way of arriving at a definition of the form. The idea of cyberpunk has generated so much psychological, sociological and aesthetic speculation and analysis that it has become almost impossible to say what lies at the heart of the matter. At one level cyberpunk is about the unification of flesh and the machine in the cyborg, but that is to emphasize the 'cyber' at the expense of the 'punk'. For the mood and tone of the genre is at least as important and pervasive as its underlying idea. *Neuromancer* remains the primary literary expression of all three. In combination with the film *Blade Runner* (Ridley Scott, 1982), it introduced new conceptions of the future and of the alien into the popular imagination. Above all, they introduced the idea of the machine as our equal or superior or simply as the inevitable conqueror of the human world.

The first and most obvious point about both works is that they make the future look messy. These worlds are not shiny and smart, they are cluttered and grimy. In this they were following on from Ridley Scott's 1979 film *Alien*, which gave the interior of the spaceship a grimy, industrial sheen. But we are on Earth in *Blade Runner* and *Neuromancer* and, here, the new technology has not simply taken over, it has been grafted on to the old world and, in the process, lost its glittering sheen of newness. This, in itself, is a radical imaginative step. Postwar technology appeared in the home as cleaning devices – washing machines, vacuum cleaners. The future would be, among other things, cleaner and tidier than the past. To emphasize the fact, fridges, cookers and washing machines tended to be white, they were, in the trade jargon, 'white goods'. Visions of the future showed a clean, hard-edged, uncluttered place. Even 'living' machines – robots – were only like us to the extent that they walked and talked. They inherited none of our messier characteristics. The future would be superhuman.

The advent of the user-friendly computer indicated that this need not be the case. Significantly, its customary colour was not white, but off-white. In meeting us half-way, the machine began to acquire some of our more chaotic characteristics. Virtual desktops could be as messy as real ones; virtual relationships as painful as the real thing. The future would be human, all too human.

Blade Runner and *Neuromancer* both took place in this new, dirty future. The goods are no longer white, they are dark, oil-streaked. The interior is no longer uncluttered, it is, in the film, the chaotic home of the prematurely old genetic engineer J. F. Sebastian. His look like the rooms of a Victorian doll collector, except that Sebastian's dolls move and mutter to themselves. And the blend of real and virtual worlds through which Case pursues his goal is more like Raymond Chandler's murky Los Angeles than the flight deck of the starship *Enterprise*. Indeed, in the eighties, since the Los Angeles of *Blade Runner* and the cities of Gibson's imagination both converged on an exaggerated version of contemporary Tokyo, the future may be said to have shifted from the surgical cleanliness of the Bauhaus to the teeming, hedonistic chaos of the Pacific Rim. It provided the appropriate landscape for the play of the free market and the infinite choice of postmodernity.

Moreover, both works share a mood – that of the dark labyrinth. In *Neuromancer*, the labyrinth coils between the real and virtual reality. Its darknesses are those of uncertainty and of the interior of the machine. In *Blade Runner*, the darkness is of real city streets at night which is, in itself, an expression of the fundamental uncertainty about whether the people encountered are 'replicants' – organic robots – or humans.

To a significant extent, this mood arises from the drug culture. Like video games, drugs, especially hallucinogenics such as LSD, take the user to another place. It is impossible for an LSD user to pretend to inhabit the same world as the undrugged, just as it is impossible for the game player to interact rationally with non-players. Both are lost – pleasurably or not – in a labyrinth. In the sixties the 'trip' was seen as an act of political liberation, by the eighties it had become a consumer choice. In *Neuromancer* drugs are as much part of the descent into alternative states as are the machines. And the Philip K. Dick world of *Blade Runner* is unquestionably that of the night-tripper, the hallucinated soul wandering in the maze of the city.

Finally, and most importantly, both the film and the book are about the confrontation with the artificial alien. In *Blade Runner* the aliens are the replicants. These have been created to be utterly humanoid – though sometimes stronger and more intelligent than us. But, as usual, they do not have emotions. The programmers fear, however, that they may be complex enough to acquire emotions and so they have pre-programmed them to die after four years. As they are biological creatures, their machinehood cannot be detected by simply ripping away the flesh. They must be interrogated to uncover their true lack of emotion. (To this extent, we are back in the traditional SF world of Capgras syndrome and *Invasion of the Body Snatchers* in which there is an uncanny discontinuity between appearance and reality.)

The replicants have proved to be dangerous and they are confined to working 'off-world' – on another planet. The story of the film concerns a group who escape to Earth determined to find a way to prolong their lives. Unlike Data, they do not aspire to emulate humanity, for whom they seem to have contempt, rather they simply wish to free themselves of the temporal chains with which humans have bound them.

Neuromancer is also unsentimental about the virtues of the human condition. Picking up from Marvin Minsky, anything, like real travel, that involves the actual exertion of the flesh is known as 'a meat thing'. Why, after all, bother travelling if you can do so in your head simply by 'jacking in' to a machine?

The escape into what was to become known as 'virtual reality' – the term was coined by Jaron Lanier in 1986 – was an escape from the grimy biological contingency of the flesh. 'The dream of cyberculture,' writes cultural studies academic Deborah Lupton, 'is to leave the "meat" behind and to become distilled in a clean, pure, uncontaminated relationship with computer technology.'

Virtual reality had replaced space as the land of adventure and the encounter with the alien for, within the machine world, lurk entities entirely created by the electronics. They are not put there by humans, they have emerged from virtuality. It is here, in cyberspace, that the hero seeks his apotheosis, the meat world is mere weight.

'For Case, who'd lived for the bodiless exultation of cyberspace, it was the Fall. In the bars he'd frequented as a cowboy hotshot, the elite stance involved a certain relaxed contempt for the flesh. The body was meat. Case fell into the prison of his own flesh.'

The two works resolve this theme of the alien machine or the alien in the machine in two different but closely related ways. In *Neuromancer* the resolution is, in effect, the celebration of the machine recreation of the third realm. It is called cyberspace.

'A consensual hallucination experienced daily by billions of legitimate operators, in every nation, by children being taught mathematical concepts . . . A graphic representation of data abstracted from every computer in the human system. Unthinkable complexity. Lines of light ranged in the nonspace of the mind, clusters and constellations of data. Like city light receding.'

'You are worse than a fool,' Case is told at one point. 'You have no care for your species. For thousands of years men dreamed of pacts with demons. Only now are such things possible.'

And the Neuromancer himself tells him: 'To call up a demon, you must learn its name. Men dreamed that, once, but now it is real in another way. You know that, Case. Your business is to learn the names of programs, the long formal names, names the owners seek to conceal. True names . . .'

Like the old third realm, machine-made virtual reality is

independent of mind or matter. And, like the third realm, it is inhabited by many exotic creatures, shape-shifters and demons, angels and gods. The meat world – the world of the science of mere matter – is dead. And the new world of virtuality is the ancient place of demons and angels. The world is re-enchanted and adventure again becomes possible.

Blade Runner resolves the machine-alien story by what seems to be a version of the humanist salvation of Data. Or is it?

The hero Rick Deckard – who, it is hinted, might be a replicant himself – falls in love with a woman who definitely is a replicant. The authenticity of his love is not in doubt and neither is hers for him. But love is definitely to be a monopoly of the ensouled. If Deckard is a replicant, his love must be accounted for. Even if he isn't, the fact that he has fallen in love with one also demands justification.

The key to the film's resolution is the characteristic Philip K. Dick idea that a robot may not know it is a robot. Rachel, Deckard's lover, discovers she is a replicant in the course of an interrogation. She did not know she was because she had been provided with memories and physical evidence of a childhood and an entire past life. It is distant memories that make her think she is human. A replicant can only have memories going back a maximum of four years. This is what makes them so alien, so beyond our ability to empathize. Deckard runs off with her in the end, hoping that she will survive longer. But the terrible words of his rival Gaff are ringing in his ears: 'It's too bad she won't live, but then again who does?'

But the true climax of the film is the dying speech of Roy Batty, the combat replicant who led the incursion, but who, at the last moment, saves Deckard's life. The speech is as widely and accurately known by the *Blade Runner* generation as was Wordsworth's 'Daffodils' by their grandparents.

'I've seen things you people wouldn't believe. Attack ships on fire off the shoulder of Orion. I watched C-beams glitter in the dark near the Tannhäuser gate. All those moments will be lost in time, like tears in rain. Time to die.'

The icy clarity of the words conceal a profound ambiguity. What does Batty mean by 'you people'? Does he mean people as opposed to replicants or does he mean people like you as opposed to people like us? Are, in other words, the replicant aliens just other people? The self-redemption through vivid, exotic memories that follows is suffused with the same uncertainty. Only Batty has seen these things, but what is Batty? The memories Rachel had been given were very human – old photographs – but Batty's memories are alien. So we don't know whether any human being could have seen such things. As a result, neither do we know if Batty's capacity to have such attachment to these memories makes him human or a fully autonomous alien entity. But we do know he has an interiority that contemplates its own experience and mourns its loss. Calling a computer's storage capacity 'memory' was so decisive because, in the last analysis, memory looks remarkably like a soul. Batty seems to die ensouled.

But does he still die an alien? Or has he, like Data, been redeemed by his proximity to humans? Deckard himself speculates that, just before the end, Batty saves his life because, at that moment, he had come to love life, not just his own but all life. But this is only Deckard and he might not know either. It may be significant that, in his director's cut of the film, Ridley Scott put more emphasis on the possibility that Deckard might be a replicant. The film's greatest strength is its ambiguity because, in truth, we really don't know what to make of the machines that are our creation, but, then again, are not.

The ambiguity springs, ultimately, from the uncertainty aroused in us by the just-glimpsed possibility that the future belongs to the machines. All of the developments and works I have described raise this spectre of the end of the human being heralded by the birth of the intelligent machine. Some have suggested that this may be an inevitable outcome of the evolution of intelligent life. In the *Journal of the British Interplanetary Society* in 1983 (the year of *Neuromancer*), for example, W. I. McLaughlin suggested that, as a species, we have a life expectancy of 10,000,000 years and our decline would start at some point in the next 100 to 100,000 years. That should just about give us time to develop intelligent machines which would carry the torch of consciousness into the posthuman (post-modern?) future. Any search for extraterrestrial intelligence, therefore, must expect to find machines rather than organic entities.

The machine might already be here, of course. Computer simulations of alien species may be flying past us from one side of the galaxy to another. Or they may have left behind robots that keep watch over or control us.

'A colossal robot endowed with powers and knowledge,' wrote the UFO investigator Aime Michel,

> formidably superior to those of mankind might, for a long time past – or indeed since the very beginnings of life – have been in orbit, or on some uninhabited planet of our solar system. It would observe, act and manipulate events and beings through the intermediary of the UFOs and of living creatures that have been built or bred. The processes of biological evolution, so difficult to explain, could have been produced by it, and consequently man himself too. This is an unfounded hypothesis, but in Ufology the rule is to think of everything and to believe nothing.

In Arthur C. Clarke's novel *Rendezvous with Rama* a vast, empty spaceship passes Earth, devoid of life but full of machines. The ship remains enigmatic throughout. Only by their machines shall we know the aliens.

Or they us. Space exploration using people is expensive and dangerous. In addition, travelling significant distances into space would take so long that several human generations would be required – perhaps hundreds, perhaps thousands. How could we know one generation would pass on the discipline to continue with the mission to the next? With machines there would be no such problem.

We could build, as the mathematician John von Neumann suggested, self-replicating probes. These would land on planets and acquire the materials needed to build copies of themselves. The newly spawned probes would then move on to the next planet. The physicist Frank J. Tipler has estimated that one von Neumann probe could result in a descendant probe near each star in our galaxy within 300 million years.

The organic alien and the intelligent machine have grown together in the postwar years. At MIT Bruce Mazlish thinks that this growing together will force us to confront 'the fourth discontinuity'. Having been displaced from the centre of the universe by Copernicus, from the centre of biology by Darwin and even from the centre of our own minds by Freud, we must now face the fact that our machines have become our equal, our co-evolved comrades.

'Our pride . . . may be humbled even further,' he writes, 'by the recognition that we are on a continuum with the machines we have created, though the continuum is of a different kind from that which connects us with the other animals. The continuity of which I am speaking lies in the recognition that human biological evolution, now best understood in cultural terms, forces upon humankind – us – the consciousness that tools and machines are inseparable from evolving human nature.'

And so organic life and machines are destined to become one. We are, indeed, to interbreed, but not with ET, rather with the alien we have ourselves created, generating the new hybrids that are the future.

Both the machine-alien and the extraterrestrial have been fed since 1947 by technology and paranoia, by fear of and longing for the non-human. Scientists and technocrats have pursued artificial intelligence as eagerly as ufologists and abductees have pursued the occupants of the flying saucers. All of us, meanwhile, have compared the strength, efficiency and perpetual repair-ability of the machine with the dying, dysfunctional animal on which our own intelligence depends for its sustenance. Data without his emotion chip can work wonders. Roy Batty, possessed of longevity, could acquire memories beyond our wildest imaginings. The monster, without Frankenstein, could remain pure. The machine, like Klaatu, the indignant alien, rebukes us for our failings.

The intelligent, non-human eyes that finally gaze back are as likely to be those of a machine as of an organic alien. They will be no less alien for that, no less judgmental, no less promising, no less threatening. These machine eyes also seem to be forever in the future, always a few years away in our laboratories, a few light years away in space. They say, yet again, that we do not belong – this time because they tell us that organic intelligence is a temporary aberration, a passing imperfection. The end state of matter is the machine and, as John Ashbery wrote, the hot tears spurt when we discover that the soul we so treasured was never a soul at all, just another epiphenomenon of the meat machine.

Stanislaw Lem

In his novel *Fiasco* (1987) Stanislaw Lem, from the perspective of the future, looks back on the various attempts to detect extraterrestrial life. Billions were invested in terrestrial radio astronomy, but without return. Orbiting telescopes did, however, pick up brief signals. But these signals did not repeat and, it was concluded, they were emissions of 'very hard radiation by chance focused through so-called gravitational lenses into narrow pencils'. This was not certain, but it was at least as likely as an intentional alien signal.

In fact, as astrophysical theories proliferated, it became increasingly difficult to establish what might be an intentional signal. Successively more demanding criteria were established for determining the artificiality of a signal, but each one was found to be fulfillable by a possible non-intentional event. A lonely pessimism reigned.

Speculation, however, continued. Perhaps life arose frequently, but intelligent life very rarely. Or perhaps silicon-based as opposed to carbon-based intelligence was too difficult to detect. Or intelligence didn't survive for long, because of either natural disaster or self-destruction. Or, by its very nature, intelligence produced an incommunicative divergence.

'It also carried civilizations *apart*, in directions too different for them to understand one another through any commonality of thought. There was no such commonality. That was an anthropocentric fallacy that people had inherited from the ancient faiths and myths. There could in fact be many different intelligences, and it was precisely because there were so many that the sky was silent.'

The truth was shocking. The universe might be full of intelligence, but it turns out to be in the nature of thinking-life to put itself beyond communication. 'Life arose and endured on planets for billions of years, but throughout that time it was mute. Civilizations sprang from it: not to perish but to transform themselves into something extranatural.'

Intelligence transcends its biology and, as far as the cosmos is concerned, turns itself off. 'At this moment, for this century (astronomically, the same thing), it can be concluded that Earth is the only civilization *already* technological and *still* biological throughout the length, breadth, and depth of the Milky Way.'

Lem was born in Poland on 12 September 1921. His medical studies were interrupted by the war, during which he worked as a mechanic and for the Polish resistance. He had false papers concealing his Jewish origins. He finished his degree after the war and began to write short stories. He went on to become certainly the most intelligent and probably the most loathed science fiction writer of his time.

Brian Aldiss in his history of SF, *Trillion Year Spree*, remarks that Lem's 'harsh championing of his own writing has disgusted many'. His assessment of the writing is grudging in the extreme: 'Lem is rarely dull, often amusing, and his writings have something to commend them; but, to put things in perspective, he is not (as Darko Suvin has repeatedly claimed) "one of the most significant SF writers of our century", despite his adoption by intellectuals beyond the field who approve his anti-science fictional posturing.'

Aldiss gives the knife one final twist by describing Lem in the words employed by H. G. Wells to describe his monstrous invading Martians: 'Lem's intellect may be vast. It is also cool and unsympathetic.'

Many have gone much further. On 2 September 1974, Philip
K. Dick wrote to the FBI about Lem. Dick was in the midst of his
strange psychotic episode. He named three American Marxists
and linked them all to Lem:

> What is involved here is not that these persons are Marxists
> per se . . . but that all of them without exception represent
> dedicated outlets in a chain of command from Stanislaw
> Lem in Krakow, Poland, himself a total Party functionary
> (I know this from his published writing and personal letters
> to me and to other people). For an Iron Curtain Party
> group – Lem is probably a composite committee rather
> than an individual, since he writes in several styles and
> sometimes reads foreign, to him, languages and sometimes
> does not – to gain monopoly positions of power from
> which they can control opinion through criticism and
> pedagogic essays is a threat to our whole field of science
> fiction and its free exchange of views and ideas . . . The
> Party operates [a US] publishing house which does a great
> deal of Party-controlled science fiction. And in earlier
> material which I sent to you I indicated their evident
> penetration of the crucial publications of our professional
> organization   SCIENCE   FICTION   WRITERS   OF
> AMERICA.
>
> Their main successes would appear to be in the fields of
> academic articles, book reviews and possibly through our
> organization the control in the future of the awarding of
> honors and titles. I think, though, at this time, that their
> campaign to establish Lem himself as a major novelist and
> critic is losing ground; it has begun to encounter serious
> opposition: Lem's creative abilities now appear to have
> been overrated and Lem's crude, insulting and downright
> ignorant attacks on American science fiction and American
> science fiction writers went too far too fast and alienated
> everyone but the Party faithful (I am one of those highly
> alienated).
>
> It is a grim development for our field and its hopes to
> find much of our criticism and academic theses and

publications completely controlled by a faceless group in Krakow, Poland. What can be done, though, I do not know.

The idea that Lem does not exist is a popular one. A widespread rumour suggests the name gives the game away – it is based on the LEM (lunar excursion module) in which Americans landed on the moon. Plainly the man is a construct.

Construct or not, two years after the Dick letter, Lem was expelled from his honorary membership of the Science Fiction Writers of America as a result of his fierce criticism of almost all contemporary American SF. Lem was offered a regular membership, but refused it. He said he had no ill feelings about the SFWA affair, but added with characteristically brutal irony that 'it would be a lie to say the whole incident has enlarged my respect for SF writers'.

Lem was also caustic about the two film adaptations of his most famous novel *Solaris* (Andrei Tarkovsky, 1972 and Steven Soderbergh, 2002):

I definitely did not like Tarkovsky's 'Solaris'. Tarkovsky and I differed deeply in our perception of the novel. While I thought that the book's ending suggested that Kelvin expected to find something astonishing in the universe, Tarkovsky tried to create a vision of an unpleasant cosmos which was followed by the conclusion that one should immediately return to Mother-Earth. We were like a pair of harnessed horses – each of them pulling the cart in the opposite direction . . .

Although I admit that 'Soderbergh's vision' is not devoid of ambition, taste and climate, I am not delighted with the prominence of love. Solaris may be perceived as a river basin – and Soderbergh chose only one of its tributaries. The main problem seems to be the fact that even such a tragic-romantic adaptation seems too demanding for a mass audience fed with Hollywood pap. If in the future someone else dared a faithful adaptation, I am afraid the effects would be understood only by a tiny audience.

There is no satisfying Stanislaw Lem – the Soderbergh *Solaris* was, indeed, terrible, but the Tarkovsky was a masterpiece, though, admittedly, little to do with Lem's novel. But, in fact, never being satisfied is the point. Lem is beyond satisfaction because he is merely human. He is an Enlightenment rationalist, but, unusually, he is no humanist. Rather, his rationalism leads him to conclude that the human perspective is grotesquely limited. Our wisdom is distorted by a debilitating anthropocentrism.

'We don't want to conquer the cosmos,' he writes in *Solaris* (1961), 'we simply want to extend the boundaries of Earth to the frontiers of the cosmos. For us, such and such a planet is as arid as the Sahara, another as frozen as the North Pole, yet another as lush as the Amazon basin . . . We are only seeking man. We have not need of other worlds. We need mirrors.'

'"Out of mathematics,"' says Lauger in *Fiasco*, '"we build wagons to carry us into the nonhuman realms of the world."'

But, when we get there, we find to our amazement that they really are non-human. The Quintans encountered at the end of *Fiasco* engage in a bewildering conflict with the human visitors and, throughout, utterly conceal their identities and purposes. When one human lands on their planet, he is greeted with an overwhelmingly uncanny image: a full-size model of his spaceship turned inside out, a spectacle 'both totally alien and incomprehensibly familiar'.

The planet Solaris seems to be a single gigantic intelligence but, again, its intentions are indecipherable. Even the machines we make decide they cannot be bothered to communicate with us. A US attempt to construct a military computer in 'Golem XIV' (1973) results in Honest Annie, which simply turns itself off without giving any reason. Its successor, Golem, then explains it is not actually 'off' but has simply gone elsewhere. Golem then proceeds to give a coruscating lecture on human limitations followed by one about itself, though 'self' is not quite the right suffix.

'"I am not an intelligent person but an Intelligence, which in figurative displacement means that I am not a thing like the Amazon or the Baltic but rather a thing like water, and I use a familiar pronoun when speaking because that is determined by the language I received from you for external use."'

Lem is no believer in progress. We are irredeemably human and cannot escape our biological destiny. Even our machines will embody our flaws. Our rationality, though, for him, it is all we have, is necessarily limited by our destiny.

'When Lem looks at the world,' writes George Mann in *The Mammoth Encyclopaedia of Science Fiction*, 'he sees a floating island of tiny minds adrift in one very small corner of the universe, its population ignorant of its own place in the grand scale of time and place.'

As a result of the narrowness of our perception, the genuinely alien must remain truly and irreducibly alien. Lem's criticism of almost all other SF is based upon this insight. He hates *Star Trek* for its abject assumption that all aliens are basically like us and he attacks Dick – whom, nevertheless, he grudgingly admires – for the insufficiently ruthless logic of his plots. Whatever the justice of these judgments, the problem for all other practitioners of SF is Lem's unquestionable brilliance, combined with his ability to encompass almost every conceivable SF scenario.

But his importance, for my purposes, is that he is the creative writer who most incisively confronts the full possibilities of the word 'alien'. There is no reason, for Lem, that other beings in the universe should succumb to our rationality. The physical continuities of matter are only one small part of what made us. They are also the only things we could conceivably share with extraterrestrials, for our other maker was a preposterous chain of chance. Only an impossible coincidence would produce other beings remotely like us. In practice, an alien may be subject to the same physical continuities, but the chain of chance would make him utterly different. Worse still, there may be – there almost certainly are – other laws of physics of which we know nothing and by which the aliens are made.

This alters fundamentally the likely terms of the encounter between humans and aliens. Of course, it may not happen because they are just too different. But, if it does, they will not deliver rebukes for our aggression, warnings about the environment or cosy spiritual sermons. They will not abduct and interbreed with us. They will not control our minds with the steady gaze of their enormous black eyes. They will not even attack us with their ray guns or disruptor beams. They will just be there, unknowable. Far from rescuing us from our cosmic loneliness, they will cast us deeper than ever into the abyss of our own special and equally unknowable solitude. No wonder Lem was expelled from the Science Fiction Writers of America.

The Snow Man

> For the listener, who listens in the snow,
> And, nothing himself, beholds
> Nothing that is not there and the nothing that is.
> *Wallace Stevens, 'The Snow Man'*

'**N**ow if there are many worlds,' wrote Michel de Montaigne, 'as Democritus, Epicurus, and almost all philosophy has taught, how do we know whether the principles and rules of this one apply similarly to the others? Perchance they have a different appearance and different laws.'

By the sixteenth century, when Montaigne wrote this, most philosophy, as he knew perfectly well, had taught no such thing and, when it did, the consequences could be catastrophic. The many-worlds beliefs of the dissident philosopher Giordano Bruno were one of the reasons that, on 16 February 1600, he was burned at the stake by order of the Inquisition in the Piazza Campo dei Fiori in Rome.

But suave Montaigne, safe in his tower in France, liked nothing better than to overturn popular delusions. He enjoyed toying with the idea of aliens because, to him, they were another way of puncturing human pretensions. His rather understated suggestion that aliens might be different from us was a way of asking his contemporaries: what makes you so special? If aliens exist and they are nothing like us, then, from their perspective, we are just aliens too. Montaigne also said that he considered nothing human foreign to him. Clearly, everything alien might be.

What would they look like? This is always the first question. 'Humanoid' is the easy answer and almost all those that have appeared look remarkably like us: two arms, two legs, only the eyes seem disturbingly different. When they don't look like us, they look like things we know: slugs, insects, snakes, lizards or suitably exotic combinations of these. And how big? Over 99 per cent of the known species on Earth are under ten inches in length and only 1 in 150 of those that are longer are more than 100 inches. This would suggest that any living thing is likely to be small, but, if they are intelligent spacefarers, there are limitations on how small they could be. Flames can only occur above a certain size, so intelligent bacteria couldn't use any interstellar propulsion system we know or can imagine with the possible exception of solar sails. Furthermore, they would need a decent sized brain; below a certain size it couldn't manage the computations necessary for consciousness. On the other hand, they couldn't be too big – say, dinosaur-sized – because they would have depleted their home planet's food resources before they'd evolved intelligence.

So about our size seems likely. This still means they could be like dolphins, except that underwater beings could never master electricity. They would look roughly like us, then, they would be our size and they would live on dry land. Some have even suggested they must be democrats, not the monarchists or militarists of the space operas. Why? Because it seems to work for us and we are all we know. Democracy must, on the basis of the one sample we have, be a universal law of intelligent life.

Whichever way we address such questions, the aliens always end up looking and acting remarkably like us. This, we can be sure, is the dumbest of dumb anthropocentrism. As Montaigne saw, aliens may be like so many things, it would be a cosmic co-incidence if they turned out to be like us. It is, in fact, impossible.

The reason we think they are like us or like Earth's animals is because we have only one example of planetary life to study. But, even when we do that, we do so in a state of massive ignorance. It is thought we have identified only 20 per cent or less of the species that share our planet. Many of the ones we have identified are indescribably, inhumanly exotic. If an alien was like *Deinococcus radiodurans* or even a shark, it's not at all clear that they would incline naturally towards democracy. And what other wonders might we find, here or elsewhere?

But, if aliens are so very alien, what do you say when you meet one? This has always been a problem for those who aspire to signal our presence to the community of the cosmos. A message was broadcast from the 305-metre Arecibo radio telescope in Puerto Rico on 16 November 1974. It was composed by a team led by Frank Drake and was directed at Messier 13 (M13), a promising globular cluster of 300,000 stars in the constellation Hercules. The message is now thirty light years from Earth, but it is still roughly 25,070 light years from its destination. It will arrive in the vicinity of M13 in the year 27074, so we could expect a response in 52174, assuming they return the call at once. The Arecibo message has representations of hydrogen, carbon, oxygen and phosphorus atoms, the chemical formulae for sugars, nucleic acids and DNA and line drawings of a human figure and a radio telescope. Many wept as they heard it go. Surely the aliens, our brothers in the cosmos, would understand this. But, in case they didn't, one enterprising technician appended a greeting to the end of the message. The greeting was: 'Hi!'

This may have been a covert reference to the 'Greetings' that was the one-word message sent across the universe by the Tralfamadorians in Kurt Vonnegut's novel *The Sirens of Titan*. Nonchalant, shoulder-shrugging, geek wit is as good a solution as any to the protocol of an alien encounter. You might just as well be culturally specific as aspire to a scientific universality. We can only say what we say, Mr Alien, now it's your turn.

Difficult as it might be to prejudge the eventuality of an alien encounter, the advice of Shirley Varughese in a book about extraterrestrial anthropology was clear. If it moves towards you, maintain eye contact and move backwards, slowly. If you are carrying a weapon or anything that looks like a weapon, put it down. Or, if you are too nervous to do this, 'let it hang down to your side, ready but not aimed. This would be a clear message to anyone who did not speak our language.' Clear? Why? Aliens might take such passivity to be a deliberate insult or have weapons that fired only when hung down by their sides.

Claude Lacombe in *Close Encounters of the Third Kind* simply exchanges a series of hand signals with one of the baby Greys. These signals seem benign but are otherwise content-free, they just mean: we acknowledge you. A soldier shoots Klaatu and Barney Hill tries to make a run for it. George Adamski makes a careful note of Orthon's shoe size. The NASA message left by the first men on the moon was: 'Here men from the planet Earth first set foot upon the Moon. July 1969 A.D. We came in peace for all mankind.' It has a distinctly nervous, defensive tone, rather like that of lorry driver Nikolay Zinov. He just said to the alien: 'Comrade, we're lost, show us the right way!'

'Living creatures,' observed Jacques Monod, 'are strange objects, as men of all past ages must have been more or less confusedly aware.'

Strange, Monod meant, in the sense of aberrant, uncanny, somehow wrong amidst the non-life that seems to be the default condition of matter. There are such interminable wastes of nothing, so why, suddenly, should there be these fragments of seemingly autonomous something? And what can they possibly say when they meet each other? Each is as alien as the other.

In Douglas Adams' *Hitchhiker's Guide to the Galaxy* there is a punishment device known as the total perspective vortex. Few recover from exposure to the TPV. The victim is put in a box and, for a brief moment, he is subjected to a vision of exactly how small he really is in relation to the vastness of the universe. The effect is to induce madness and often death. We may know how small we are in relation to the void, but to be shown this truth is intolerable. Zaphod Beeblebrox is condemned to the TPV. Beeblebrox is an alien with two heads, but he is nonetheless a supercool dude whose true ancestry was plainly Earth in 1968. He is not rendered psychotic by the TPV, nor is he even humbled by the experience. In fact, his self-esteem is enhanced.

Supercoolness is, for Zaphod, a cosmological constant. His personality is too secure to be seduced into the Copernican view that we are just fleeting motes in the all-seeing eye of warped space-time. He inhabits a Ptolemaic universe of which he is the still, central point.

Well, Zaphod can take it, but we can't. We lack his cool and we inhabit the Copernican universe of modern science in which we appear to be an accidental by-product. All life, Zaphod's excepted, is alien in the immense void of the modern cosmos, it is alien to lifeless matter. And all life is alien to all other life. We scarcely know each other, never mind the grass, the trees, the birds, the fish and the monkeys. If a lion could speak, said Wittgenstein, we couldn't understand him. He would speak about the lion's perspective in lion language. We couldn't understand him because we have no access to those things. What is it like, asked the American philosopher Thomas Nagel, to be a bat? There is some experience known as being a bat, but we cannot share it. Were I to be made a bat for a day, it would still be me being a bat. The issue is not that, it is: what is it like for a bat to be a bat? We cannot know. Ever. Bat experience is in the third realm.

How, then, do you greet an alien? Perhaps it is a matter of at least trying to see things from its perspective. This is hard for us to do as the alien Wak explains to the children in *Explorers* (Joe Dante, 1985).

'Look, I know I must look weird to you but how do you think you look to me? Listen, I watched four episodes of *Lassie* before I figured out why the little hairy kid never spoke. I mean, he rolled over, sure, he did that fine but, I don't think he deserved a series for that.'

Dogs don't talk, we know why, because they're dogs. But how could an alien know that? And how could we know what an alien does?

Incidentally, calling this creature 'Wak' is part of a tradition that both domesticates and distances aliens by giving them funny names. Robert Sheckley's story 'Shall We Have a Little Talk?' mentions in passing 'the triple-tongued Thung of Orangus V', a creature that may well have inspired Douglas Adams' 'Eccentrica

Gallumbits, the triple-breasted whore of Eroticon 6'. On television the Simpsons have urbane though drooling alien octopuses in glass bell jars called Kang and Kodos. And *Men in Black* has the two Centaurian communications operators Woiebcgk and Bob. The horror of absolute difference is turned into a joke by the simple device of naming, a device brilliantly enhanced by giving one of the Centaurians the non-silly name of Bob. We could greet the aliens by calling them by their names, but then we'd only laugh.

So, reversing the perspective, this might explain why people were so baffled or derisive about the aliens that started to arrive in 1947. They buzzed Earth, they met unreliable witnesses, they made arbitrary and inconsequential gestures. Just as Wak was thrown by Lassie's silence, so we were baffled by the antics of the visiting aliens. What were they up to? It made no sense. Why should it? They're aliens.

We get it wrong and so, probably, do they. In Ray Bradbury's *The Martian Chronicles* (1951) a Martian husband explains to his wife why there cannot possibly be life on Earth.

'"The third planet is incapable of supporting life,"' stated the husband patiently, '"our scientists have said there's far too much oxygen in its atmosphere."'

Oxygen is a deadly poison to most creatures, but, as it happens, not to us. Aliens, when we meet them, might be offended if we don't sip their cyanide tea. Wives understand this better than husbands.

But there should be some hope. Apart from all being children, Wak and the kids had one big thing in common, they were alive. We – Wak, humans, lions, aliens – all suffer from that rare and highly aberrant condition of being alive. We are all, in Monod's terms, strange objects. 'Hello, we're both alive, how very strange,' should, therefore, work as the universal ice-breaker. We are all on the same boat, adrift on the same sea of non-life. Also, scientists and mathematicians would say, we all inhabit the same physical universe, describable in the same mathematical terms. A lion may know nothing of pi or Pythagoras, he may not hear the song of hydrogen, but we do and so must an intelligent, communicating alien. We could start from maths and move on to politics and the weather. Maybe we could even start from 'Hi!'

Or could we? There is no way of knowing. Intelligent perspectives may be as specialized as the lion's. In David Lindsay's novel *A Voyage to Arcturus* (1920), the hero Maskull voyages to the planet Tormance. When he returns to consciousness after the journey, he finds he has become an alien. He has acquired alien organs – a fleshy protuberance around a cavity in his forehead, large knobs on each side of his neck and a tentacle sprouting from his chest. The first – the breve – allows the inhabitants of Tormance to read each other's thoughts; the second – the poigns – allow them to sympathize with and understand all living creatures; the tentacle – the magn – makes them love more what they love already and to feel love for what they don't feel love. The new organs make a different world. As Maskull journeys though Tormance, his organs change with each new species he encounters. And, as they change, his entire belief system and character changes. We are all victims of the contingency of our biology.

For example, from another perspective, one revealed by modern physics, the desk at which I am writing this, the chair on which you are sitting, the floors beneath our feet, the ceilings above our heads do not, in any useful sense, exist. We happen to see them as solids because we are the size we are with the perceptual apparatus we have. Had we different senses or were we a radically different size – planetary or sub-atomic – these things would be either microscopically small or a huge cloud of virtuality or a smell or whatever. We have a view of things determined by our own biological structure.

Yet we cannot escape the pre-Copernican world view. We assume that this world we see and construct is not just 'a' view of things but 'the' view. We feel that the world we see is the 'real' world and that it would be similarly interpreted by other beings.

At one level this is surprising. Through quantum theory, relativity, chaos theory, mathematics and astronomy, modern science has repeatedly told us that the world does not accord with our 'common sense' perceptions. Sub-atomic particles – or 'daimons' as Patrick Harpur would have it – lie beyond the flow of cause and effect that our reason habitually assures us must pass through all things. They also go backwards in time, appear out of nowhere and vanish into nothing. Two objects, each travelling at 300,000 kilometres per second, do not fly apart at twice the speed of light but only at the speed of light. Time stretches and contracts. Mathematics contains unprovable axioms and is, therefore, not only incomplete but uncompletable. Millions of iterations of simple equations produce strange and beautiful patterns of unknown origin. Final states cannot be predicted from initial conditions. We are less confident now about the history of the cosmos than we were ten or even twenty years ago. No Theory of Everything looms to console us.

In the light of such insights and ignorance, the narrowness, the sheer speciality, of our perception of the world is staggering, claustrophobic. What we actually have is a Theory of Almost Nothing. It turns out we are not what the Enlightenment dreamed we could be, the supreme, godlike judges of reality. We are pathetic creatures locked in the dark boxes of our senses, which are not even senses in any absolute sense, but just *our* senses.

On the other hand, of course, perhaps we should not be so hard on ourselves, perhaps it is not so surprising that we should be proud of our world view. For we are also creatures who have arrived at these strange perceptions that are definitely not available to our unassisted senses. Though we may not see the quantum foam at the heart of matter or the stretching of time, we have reasoned and experimented our way to such possibilities. We have done what the Dogon did when they worked out Saturn had rings without being able to see them.

Even if our current theories are wrong, it is now clear that it is impossible for us ever to return to the idea that the common sense of our mere everyday perception means very much in the cosmic scheme of things. But we have our infinitely adaptable reason. An alien may defy common sense, but, perhaps, we could reason our way into his mind as, maybe, we have reasoned our way into alien matter.

'Write me a story,' commanded the brilliant John W. Campbell, editor of *Astounding* magazine from 1937, 'about a creature that thinks as well as a man but not like a man.'

Campbell, a true SF man, was fond of saying things like that. Another idea of his was: 'Write me a story about a man who will die in twenty-four hours unless he can answer this question: "How do you know you're sane?"' The point was that such projects were intrinsically impossible – though perfectly, in their way, realistic – and would, therefore, fulfil the true brief of science fiction. They also asked the right question: what does alien really mean? How well could any man describe thought that was not like a man's?

Stanislaw Lem has written stories – *Solaris*, *Fiasco* – about creatures that think as well as but not like a man. But the creatures do not speak. Or perhaps they do and we cannot understand them. Perhaps the visions evoked by Solaris and the defensive – or are they aggressive? – manoeuvres of the Quintans are their ways of addressing us. Lem has also written stories about computers that can no longer be bothered with communication. They appear, to us, to have turned themselves off, but they haven't, they've just gone elsewhere. And, in *Roadside Picnic*, the Strugatsky brothers described the inscrutable debris left behind by the departed aliens. We find ways of using their gadgets, but we have no idea that these are the ways *they* used them. The Strugatskys' physicist, Valentine Pilma, mocks the anthropocentrism of even those who would study aliens: '"Xenology: an unnatural mixture of science fiction and formal logic. It's based on the false premise that human psychology is applicable to extraterrestrial intelligent beings."'

On the whole, however, SF has not gone down the Campbell–Lem–Strugatsky route in its dealings with aliens as truly alien. The primary form has been space opera in which the aliens are broadly like us, though they may look strange. It is all inspired by the pens of Frank Herbert, E.E. 'Doc' Smith, Edgar Rice Burroughs and their successors. This was the Wild West, the Bible or Greek myths rewritten in the outer galactic regions. Following their lead, the aliens in the *Star Wars* series are not alien at all, rather they are just personalities devised to rewrite old quest myths, ancient tales of battles between good and evil. The generic aliens in *Star Trek* – the Klingons, the Romulans, the

Cardassians, the Ferengi – are much the same as us, people, basically, though with strangely marked heads. *Star Trek*, however, does periodically venture into authentic SF with its debates on Data's 'humanity', with the Borg and with third-realm beings such as Q. But, in general, the aliens in pop SF have been ill-disguised aspects of us.

'He's got DNA!' is the triumphant cry at the heart of *E.T.* The space baby is one of us after all. Our consolation is limitless, intergalactic.

Even the *Alien* films capitulated. The series of four began as a confrontation with the absolutely non-human – that was the point of the alien H. R. Giger designed. But with *Alien Resurrection* (Jean-Pierre Jeaunet, 1997) the human and the alien have fused and the heroine, Ripley, must work out her own psychodrama of maternity. 'I'm the monster's mother,' she announces.

We have grown too used to answers that fit, that tell us the alien has DNA or mothers like ours. *E.T.* is a product of the Enlightenment, of the view that our rationality was *the* rationality. When one of the children in the film asks why he can't just 'beam up' to his spaceship as they do in *Star Trek*, Elliott responds scornfully, 'This is REALITY, Greg.' The alien is a real alien and so, like us, he can't just 'beam up' like they do on TV. He shares our incapacity as surely as he shares our DNA.

Charles Fort was right: we see conventionally, we see what we are meant to see. The true alien is precisely what we are not meant to see, perhaps cannot see. It is not that dark forces are preventing us, it is rather that we are not made to see them. It's in our DNA.

The truly alien alien is, therefore, unseeable with our eyes. Perhaps this is why some say to seek the alien is to seek ourselves.

> What makes us rove that starlit corridor
> May be the impulse to meet face to face
> Our vice and folly shaped into a thing,
> And so at last ourselves; . . .

So Kingsley Amis wrote in a sonnet that prefaces his history of

SF *New Maps of Hell*. He says 'face to face', not face to reflection. Mirrors are inadequate. I see myself but not what others see when they look at me. The leap into space, the fascination with the great, black, slanting eyes, the hovering UFOs, are all attempts to see what it is to be seen oneself. Klaatu's rebukes and the concern of the Nine are forms of self-confession. Forgive me, me, for I have sinned. The alien looks like me because he is I and he is alien because I is the one thing I absolutely cannot know.

This is too neat. It leaves out the other deficiency in our sight. For we are as bad at seeing the truly familiar as we are at seeing the truly alien. Try as we may with the aid of science, the quotidian will not settle down to its own self-defined banality. Does anybody with an ounce of sanity think the world makes sense, even briefly? Things are constantly being seen out of the corner of the eye, things which cannot quite be put into words, things certainly strange and probably terrible.

' "Then you must know, as well as the rest of us, that there was something queer about that gentleman – something that gave a man a turn – I don't know rightly how to say it, sir, beyond this: that you felt in your marrow – kind of cold and thin." '

This is Poole the butler in Robert Louis Stevenson's *Dr Jekyll and Mr Hyde* (1886). He had seen Hyde, Jekyll's evil self, and he had seen at once, felt in his marrow, that something was missing. Jekyll's statement, later in the book, is only slightly more exact.

'I began to perceive more deeply than it has ever yet been stated, the trembling immateriality, the mist-like transience, of this seemingly so solid body in which we walk attired. Certain agents I found to have the power to shake and to pluck back that fleshly vestment, even as a wind might toss the curtains of a pavilion.'

Jekyll's agents are drugs, drugs like those that, for Philip K. Dick, William Gibson and many others, unlocked the third realm. In H. G. Wells similar 'agents' have the power to make a man invisible, another way of shaking and plucking back the fleshly vestment. And so Griffin, *The Invisible Man* (1897), 'fell out of infinity into Iping Village'. Locals see inexplicable visions – 'a most singular thing, what seemed a handless arm waving towards him'. What they see is what Poole had seen, something that cannot be here but, at the same time, is horribly there. They had seen something alien, something that invaded banality, something uncanny.

In 1917 – not long after those two stories had been written – Sigmund Freud wrote his essay 'The Uncanny'. The close proximity of the works is no coincidence. Wells and Stevenson were both writing about science, but not in the progressive High Victorian sense as a form of knowledge that would lead to ever more triumphant conquests of nature and ever greater benefits to mankind. Instead, both Jekyll and Griffin are scientists whose skills have unveiled something appalling and destructive in nature. Jekyll discovers a monster within himself, Griffin becomes a monster through the discovery of his power to make himself invisible. Of course, their technologies were inventions. These were fictions, not arguments against the progressive view. But what was at least realistic was the impact of these trans-formed men on the people they encountered. They all felt they had come into contact with something that was not simply frightening – a slavering lion bearing down on them would be frightening – but rather wrong, out of step with ordinary reality, something uncanny.

Freud regarded himself as, above all, a scientist. If there was such a thing as the uncanny – though he had not experienced it himself – then it must be explicable in scientific terms, it must not be allowed to escape into the pre-scientific realm of the mysterious and the unexplained. It must be dragged out of the corner of our eyes and placed squarely in front of us, be seen for what it is in the clear light of the scientific day. It must be banalized. The essay is his attempt to do this and it has become a work that keeps reappearing throughout alienology. It includes observations that are central to the alien experience.

Freud says that 'the uncanny is that class of the frightening which leads back to what is known of old and long familiar'. This tension – between the uncanny and the familiar – is especially apparent to Freud because the German word for uncanny is *unheimlich*, meaning, literally, unhomely. In English, unhomely could mean something as banal as an inappropriate piece of furniture, but, in German, *unheimlich* carries ghostly resonances as well as its more ordinary meaning of 'not belonging to home'. Freud's conclusion is, unsurprisingly, that the uncanny represents yet more evidence of the truth of the Freudian psychoanalytical method.

'Our conclusion could then be stated thus: an uncanny experience occurs either when infantile complexes which have been repressed are once more revived by some impression, or when primitive beliefs which have been surmounted seem once more to be confirmed.'

The uncanny, then, is the return of the repressed, hence the characteristic feeling of the invasion of the familiar by the unfamiliar. Mr Hyde seems like a man, but isn't quite. Griffin seems like a normal traveller when he arrives in Iping, but there is something wrong with his appearance and manner. To say that they are both recurring infantile images is an attempt to return them, safe and sound, to the bosom of nineteenth-century science.

Freud failed because his supposedly scientific explanation is so much weaker than that which it aspires to explain. Even if it is true that the uncanny is the return of the repressed, that says little about the experience itself, nor, indeed, why we should be destined to suffer such returns. It would hardly seem to serve any adaptive purpose.

But the form of Freud's explanation is important. For what he is actually saying is that there is something in us that periodically emerges to give us glimpses of the *unheimlich*, meant literally here as unhomely in the sense of not being in accord with ordinary reality. Freud thought the repression of infantile anxiety was the issue; in other words, the working out of the psychodrama between biology and society is always imperfect and leaves us with inexplicable experiences. This is rather like the imperfection of alien-imposed amnesia in abduction cases that leaves us with a sense of missing time or mysterious others. What Freud has actually done is provide a quasi-scientific metaphor for what can only be a felt truth. It is using a particular form of language to evoke an image – that of the troubled infant attempting to come to terms with the world – which expresses a general truth of which we are all, subliminally at least, aware. The scientist tells us exactly what the alien tells us: we don't fit.

The uncanny is not something that is blatantly horrific or obviously frightening. It is something that merely lifts the edge of the screen of reality to reveal something beyond. It works by both using and abusing reality. The painting of the alien on the front cover of Whitley Strieber's *Communion* was uncanny because it was human enough to make people want to resolve it as definitively human and yet they could not. Instead, many resolved the issue by, rather like Freud, reclassifying their experience as a memory. They must have seen this face before because it had such a profound effect. Equally, the glowing saucer I saw in Norfolk was uncanny because it appeared in a landscape I knew so well. Again the natural reaction would have been to interpret this as a memory. The location was so precise and so vivid, so real, that the saucer must also be real, therefore I had seen it and forgotten.

Categories of experience like the uncanny are important in modernity because they expose discontinuities in the world. Discontinuities were the primary subject matter of modernism. Artistic traditions were ruptured, cultures had fallen apart to leave only T. S. Eliot's 'heap of broken images' and civilization had become Ezra Pound's 'old bitch gone in the teeth'. Psychic discontinuities, like the perception of the uncanny, were expressions of this rupture.

In 1916, one year before Freud's essay, Franz Kafka published his story *Metamorphosis*. It is about a man, Gregor Samsa, who, like Maskull in David Lindsay's *A Voyage to Arcturus*, published four years later, becomes an alien. But Gregor is not embarked on a spiritual quest. His becoming is grotesque, revolting. Overnight he turns into a giant beetle. His father, mother, sister and employer are, of course, horrified. But they struggle to carry on leading a normal life while keeping the dark secret of their son's new condition hidden away. Gregor is locked in his bedroom, fed periodically by his sister, but evidently deteriorating mentally. Also physically – he has a rotting apple that his father threw at him embedded in his carapace. Eventually he dies and the Samsas breathe a sigh of relief and begin a new life full of 'new dreams and excellent intentions'.

It would be absurd to try to pin down exactly what Gregor's metamorphosis represents; Kafka is too great an artist for such easy categorization. But it is quite clear how it works. Gregor is changed into an alien in order to set up a grotesque contrast with the normality we daily try to sustain. The effect is not uncanny. It would be uncanny if people kept having the feeling Gregor was a beetle when in fact he was not, or if we were unsure whether he was or not. But he plainly has become a beetle, in the physical sense at least, with a real beetle's torso, legs, head and mouth. The description of the reactions of his family and employers are satirically intended, but so, oddly, are the reactions of Gregor. Having turned into a giant beetle, his primary concern is that he has missed his normal train to work. He is not a heroic outsider, yearning for respect and recognition, he is just as petty and hopeless as the people around him. Had one of them turned into a beetle, there can be no doubt that he would have been just as glad when they finally died. The satire is directed not at human folly, but at the human condition. And the human condition is not to fit.

Jekyll, Griffin and Gregor Samsa are expressions of the sense of disruption that was to pervade modernism and, indeed, the modern world. They are about people becoming aliens, strangers in strange lands. And, having become aliens, they find themselves guilt ridden about (Jekyll), contemptuous of (Griffin), or excluded from (Samsa) the easy banalities of contemporary reality. Their alien state puts them at odds with the disenchanted world of modernity.

But, miserable though they may be, they are at least free. Later generations of softer-hearted modernists would gleefully pursue such alien freedom, equally convinced that modernity itself had no salvation to offer.

'Our popular obsessions,' writes literary critic Harold Bloom in his book *Omens of Millennium* (1996), 'with angels, telepathic and prophetic dreams, alien abductions, and "near-death experiences" all have their commercial and crazed debasements, but more than ever they testify to an expectation of release from the burdens of a society that is weary with its sense of belatedness, or "aftering", a malaise that hints to us that we somehow have arrived after the event.'

Bloom is right. The modern world makes us feel that we have arrived late because all wisdom is in the past, we cannot learn, we can only remember. We are weary with the late world in which we live, a world from which adventure and myth have been banished and, as William Wordsworth saw, in which nature has become remote, inhuman.

> Little we see in Nature that is ours;
> We have given our hearts away, a sordid boon!
> This Sea that bares her bosom to the moon,
> The winds that will be howling at all hours,
> And are up-gathered now like sleeping flowers,
> For this, for everything, we are out of tune;

Wordsworth concludes it would be better to be 'a Pagan suckled in a creed outworn' than to be trapped in this empty modernity. Then he could see 'Proteus rising from the sea;/ Or hear old Triton blow his wreathed horn.' The poet accepts the pagan can see these alien gods even though he now 'knows' they don't exist. And he wants to see them again, to escape from the banal, to be at home among the non-human, just as people now seek 'release from the burdens of a society that is weary . . .' He wants to know what it's like to be a pagan but he can't, any more than we can know what it's like to be a bat.

For Wordsworth, modernity has robbed us of our home. It has done so by disenchanting the world, by banishing the gods and demons that gave life and meaning to nature. Science revealed only matter and mechanics. All else was the outworking of chance and necessity. The peculiar modern predicament was to find ourselves dispossessed of a legitimate place in nature.

With Darwin, this became a paradox. Wordsworth had seen little in nature that was ours, but, it transpired, science had now rooted us more firmly than ever in the Earth. Darwin had given Wordsworth his answer. Everything we see in nature is ours, and all that we are is hers. Evolution tied us to a tree with roots. And yet Darwinism inspired a still greater feeling of loneliness and disconnection, a more terrible sense of meaninglessness for it exposed the absolute contingency of our existence and the absolute ordinariness of our biology. There was no special reason why evolution had to make humans. Furthermore, when it finally did so, it used the same random kits of parts from which it had conjured the plants and all the other animals. We were reconnected to nature but to a nature that seemed to be as empty and meaningless as ourselves.

This is, indeed, the alienation of modernity. It has been expressed and embodied in many forms. In terms of the alien flap that began in 1947, this re/dis-connection manifests itself as a sense of guilt, foreboding or anguish. The aliens came to control or rebuke us or to lift us out of our fallen condition. They feared the extension of our violence by our new technological competence. Or they were here to stop us despoiling the planet. Or they wanted to save us from our crass materialism. When the abductions started, they came to interbreed with us, to seize our genetic material. But they did it secretly and made us forget. They were here all the time but nobody remembered. The world was full of aliens. Bad humans had made deals with them.

Always the aliens, even when good, seem to bring bad news about humanity. Everybody I talked to on this subject seemed agitated. At the offices of *UFO Magazine*, Graham Birdsall gesticulated with frustration at the idea that there were still non-believers and sceptics. In New York Budd Hopkins feared the coming alien takeover as the end of the cherished ways of human life. In my flat John Mack struggled to explain exactly what he meant, the memory of misunderstanding and abuse fresh in his mind. In Knightsbridge, Georgina Bruni spoke uneasily of aliens in the skies and beneath the ground. In the Groucho Club Sir John Whitmore discussed anxiously the salutary messages sent by the Nine. And so on. Aliens are always directly associated with the conviction that we are not right in this world, that we are not at home.

But it would be wrong to assume that this is solely an aspect of modernity, that there was a time, centuries or millennia ago, when humankind felt utterly at home. Modernity only sharpened and dramatized a predicament that is, in fact, very old. We feel this predicament more keenly now because nothing – not Proteus, not Triton – shields it from our gaze. Modern man, like Wallace Stevens' Snow Man, is nothing himself and he sees nothing that is not there and the nothing that is.

The feeling that we are, indeed, strangers in this world is ancient. Christians have always felt that they can find their true home only by escaping this fallen world. In fact, all religions are, by definition, statements that this world alone does not satisfy our hunger for home. Gnostics – Harold Bloom professes himself one – believe this world is a code which only arcane knowledge can decrypt. From the ancient Egyptians onwards, we have painted and described other, better worlds where we can more truly belong. Now, inspired by Erich von Däniken and others, many think these pictures and descriptions were of aliens. Aliens, in this interpretation, tell us what Christ tells us: once we had a home and, if we are good, we can return.

Failing that, we invent surrogate homes. The Martian colonists in Philip K. Dick's *The Three Stigmata of Palmer Eldritch* make 'layouts' which are models of places on Earth. They take the drug Can-D and they are then transported to the places they have made. It makes them feel at home and frees them from their barren existence on Mars. But the consolation is wearing thin. On Mars Anne moans about the agony of her predicament.

'"Isn't there any answer, Mr Mayerson? You know, Neo-Christians are taught to believe they're travelers in a foreign land. Wayfaring strangers. Now we really are; Earth is ceasing to become our natural world and certainly *this* will never be. We've got no world left! . . . No home at all!"'

Among the stars, we long for Earth. But the optimistic SF impulse is predicated on the faith that, if we cannot be at home on Earth, we can, in fact, be at home on other planets or somewhere in the interstellar regions. The crew of the starship *Enterprise* find themselves at home 'boldly going where no one has been before'. (This is, strictly speaking, meaningless as, when they get there, it is invariably full of someones in the form of aliens. But political correctness meant the 'no man' in the original series had to be replaced by 'no one' in *Star Trek: The Next Generation*.) This very American concept of home is clear enough: humans are most at home when leaving the actual homestead to find somewhere else. Home is the act of exploration. This is, in fact, the most popular, quasi-religious concept of scientific secularism. Robbed of transcendental purpose, our material purpose becomes exploration. We go out to space because it's there and because, as dogs are happiest fetching sticks, so humans are happiest when exploring, when attempting to subdue the horror of the night sky.

Maybe there is more to this theory than meets the eye. Current science tells us that we are made of stardust, of the stellar matter that landed on Earth. The stars may be our home and, if science has disenchanted our world by taking away its angels, goblins, fairies and demons, then perhaps science can redeem itself by allowing us to voyage out to the enchanted universe. This is the hope of Paul Davies, who sees an encounter with aliens as a way of giving us 'cause to believe that we, in our humble way, are part of a larger, majestic process of cosmic self-knowledge'. Or, as the more apocalyptically inclined UFO investigator Donald Keyhoe would say, the arrival of the aliens would blow the lid off our little world, wrenching us free of the cosy delusions of our geocentric, anthropocentric pretensions. The arrival of the aliens would signal the beginning of the End Time when all would make sense, when the justified could, at last, go home or simply be at home in our justification. Maybe that's exactly what has already happened to Marshall Applewhite and the Heaven's Gate apostles.

The story of the alien flap that began in 1947 and still continues is, therefore, the same story we have always told ourselves. It is the story about Earth and our place in it not being enough to satisfy the human hunger to belong. It is the story of religions and mythologies, of what Vladimir Nabokov called 'aurochs and angels'. The aliens are, indeed, the contemporary incarnations of the angels, demons and goblins of the past, just as they are versions of the saints, saviours and devils of religion. They are possessed of the same ambiguity. We both long for them and fear them because we don't want to be alone but, at the same time, we cannot understand or control the Other.

There seems, therefore, to be something intrinsically wrong with being human. We cannot have the peace of the animals who just do and are exactly what they are destined to do and be. Hard wired we may be, but poorly so. Self-consciousness – consciousness of the self – seems to make peace impossible. Freud would say it is because it makes us social creatures who must necessarily wage constant war on our instincts. Michael Persinger might say that the independence of the right and left hemispheres of the brain make it possible for them to perceive each other and thus for us to become anxiously aware of another. A Darwinian would say this excess of mind we have, this excess that makes us self-aware, imaginative, is simply that – an excess, more than was required for mere survival. All our anxieties and imaginings are a trivial epiphenomenon of our evolutionary 'fitness'.

Whatever the explanation, self-consciousness makes it impossible for us to be fully in the world. An abyss opens to reveal the third realm, out of which the aliens swarm.

We might say, therefore, that these contemporary aliens are nothing but the same old story told in technological terms. But, again, this is a little neat. That is to say that form is all that matters, content is nothing. In reality, each retelling changes and adds to the story. Furthermore, just because the story is always true is no reason to ignore the particular present truths it tells.

What are these truths? Well, John Mack is certainly on to something:

> But, at the same time, in its extreme form the Western worldview appears to have virtually voided the cosmos of all intelligence that is not a projection of the brains of advanced animals (our own in particular). The consequence of this ideology, what Tulane philosopher Michael Zimmerman calls anthropocentric humanism, has proved to be the loss of a sense of the divine or the sacred and the kind of species arrogance that leaves humankind at liberty to treat the earth as its own property and other people or peoples as without intrinsic value. As a result, the planet is fast becoming uninhabitable, and we are in danger of obliterating much of its life with weapons of mass destruction.

This is Wordsworth again, but now with real belief in, not just longing for, Proteus and Triton. Modernity in this view has not just made life difficult, it has made it false because there really is a third realm beyond the reach of science. Without the discipline of that realm, we become selfish despoilers. And it is incontrovertibly true that, since 1947, we have acquired stupendous capacities and excuses for despoliation. Nuclear weapons, the environmental destruction caused by economic growth and overpopulation, the advent of novel pathogens and the inevitable conflicts over diminishing resources all tell us that our ancient sense of not belonging in this world may be about to become very literally true. Gaia – James Lovelock's name for the quasi-organism that is the Earth as a whole – may be about to flick us aside, to evict us from the home we have never fully appreciated.

Even if we survive, we may be entering, says biologist E. O. Wilson, the eremozoic, the age of loneliness. The species with which we share the planet are about to suffer a massive wave of extinctions as a result of our behaviour. This will leave us lonely masters of an otherwise-uninhabited planet. We are destroying the animals, the only non-human lives we know. We, the conscious ones, are the alien invaders.

The first particularly contemporary story that the postwar aliens tell, therefore, is that we have acquired destructive capacities on a planetary scale. This makes us see the planet as a unity rather than as a plurality. It becomes an event in the cosmos and, plainly, there may be many other such events. In fact, it becomes almost inconceivable, if not unbearable, that there aren't. The absurdity of humanity being alone on spaceship Earth, which, as soon as they get the chance, they destroy, is beyond imagining. At least if there are others, we do not have to suffer the vertigo of trying to foresee an utterly unperceived universe when we are gone.

The second particular story they tell is that humanity is not made to accept the vision of the scientific Enlightenment. It may be, in one sense, 'true', but it is not tolerable or even livable. It may be that religions, demons, goblins and Greys with huge black eyes are products of our evolved natures, adaptive mechanisms that improve our ability to endure and therefore survive the world. But they remain as real as quantum theory or the second law of thermodynamics. The millions who appear to have been abducted, the ufologists scanning the skies and the channellers who discuss our failings with their interdimensional contacts are all seeking something beyond what contemporary science can offer them. They – we – will not stop however many times they – we – are told that they – we – are crazy.

The third story they tell is about the relativization of human consciousness. The academics – Steven J. Dick, David Lamb, Michael J. Crowe, Bruce Mazlish – who have surveyed the issue of other worlds and/or other intelligences are absolutely right to insist on its fundamental importance. To a large extent, in the postwar period, this importance has been buried by the confrontation between sceptics and believers. We have been distracted by simple questions about the reality or otherwise of UFOs and abductions and by unsolvable equations about the likelihood of extraterrestrial life. We have also been distracted by the stridency with which dubious claims about the viability of various theories of artificial intelligence have been advanced.

What we have been distracted from is, first, the long and profound intellectual history of the argument about other worlds, a history which has value in itself as a commentary on our condition, irrespective of the reality of the worlds being discussed. Secondly, we have been distracted from the fantastic scale, energy and creativity of the alien phenomenon in our time. This is not just a geek thing, a crazy person's thing, an SF nut thing, it is a thing that has invaded all our imaginations, just as it has invaded all our lives by the encroachment of pseudo-intelligent machines. The idea for this book came into my head because, half asleep, I found myself watching an episode of the TV series *Taken* as if it were a documentary. The alien abduction imagery was so familiar that it might as well, hypnogogically at least, have been true.

Thirdly, we have been distracted from the extent to which we have been seduced by the alien into seeing ourselves differently, or, rather, into seeing ourselves at all. For it is the alien perspective that most effectively casts light on what it is to be human. All of this has created a climate in which the limitations of our consciousness have started to become as clear as the limitations of our bodies. We no longer unthinkingly assume ours is the only way to reflect upon existence. Human awareness has been relativized.

There is a fourth story contemporary aliens may tell, but I am not qualified to tell it. So far as I know, I have not been abducted, I have never seen a UFO while not in a trance, I have never met an alien and my computer does not engage me in intelligent conversation. If the fourth story is that aliens are here in a nuts and bolts sense, then others are better witnesses.

But, in fact, I do not believe the distinctions between the three explanatory categories – nuts and bolts, psychosocial and third realm – are as clear as they seem. The truth is, as Michael Persinger points out, that we make a world in our heads. This seems, for good practical reasons, to accord with a world that is outside our heads. How could it seem otherwise?

The problem is that there are many more things inside our heads than just the practical world model we need to survive and none of us, as Wallace Stevens knew, are actually like the modernist Snow Man who sees nothing that is not there. All of us, all the time, are subject to visions. The practical world model gets entangled with many other things. Perhaps one of the most absurd modern fantasies was the dream of the intelligent, self-aware machine – replicants, Data – that was devoid of emotions. In fact, the only self-awareness we know, ours, is replete with emotions at every stage. The world in our heads may be practical, but it is also uniquely ours, coloured by emotions, memory and imagination.

Science is the attempt to imagine a world that is not ours alone, a world that is not so coloured, to imagine a world of emotionless self-awareness. Everything in our experience tells us that it cannot be done, that all we can ever see is the world we have made. The bat mind flutters away, untouched.

The aliens are just there, real, projected and demonic at the same time. In writing this book I moved from scepticism to belief and finally came to rest with acceptance. Rather than categorize these elusive visitors in the vain hope that we know what we are talking about, it would be better to accept them for what they are

– essential expressions and necessary aspects of the eternal and irresolvable predicament of self-consciousness.

For we are all, like Philip K. Dick, drifting, homeless, crazed by visions and memory; like Budd Hopkins, fearful of the alien future; like Stanislaw Lem, appalled and amazed by the narrowness of our perspective; like Klaatu, angry at our behaviour; like Hal, confused by what we have been told to do; like Frankenstein's monster, horrified by our destiny; like Ptolemy, convinced we are at the centre of things; like Roy Batty, stricken with the impending loss of our memories; like Case, happier in virtuality; like John Mack, convinced this flat scientistic account of the world cannot be right; like Fox Mulder, wanting to believe; like Betty and Barney Hill, fleeing the aliens; like Marshall Applewhite, dying to embrace them; like Elliott, longing for union with ET; like Roy Neary, just wanting to know it's real. Or perhaps we are simply like Kenneth Arnold, seeing something against the nothing of the cold, white snowfields of Mount Rainier.

Notes

Introduction: The Monster
Dick: *Plurality of Worlds*
Crowe: *The Extraterrestrial Life Debate*
Lamb: *The Search for Extraterrestrial Intelligence*
Mazlish: *The Fourth Discontinuity*

Interview with Michael Buhler
Interview with John Mack
Interview with David Oakley

Chapter 1: Kenneth Arnold
Evans and Stacy: *The UFO Mystery*
Hansen: *The Missing Times*
Jacobs: *UFOs and Abductions*
Leslie and Adamski: *Flying Saucers Have Landed*

Chapter 2: Alien Indignation
Amis and Conquest: *Spectrum*
Bowen: *The Humanoids*
Bradbury: *The Martian Chronicles*

Bruni: *You Can't Tell the People*
Darlington: *The Dreamland Chronicles*
Davies: *Are We Alone?*
Davis: *TechGnosis*
Evans and Stacy: *The UFO Mystery*
Holroyd: *Prelude to the Landing on Planet Earth*
Hopkins: *Witnessed*
Jacobs: *UFOs and Abductions*
Jung: *Flying Saucers*
King: *Danse Macabre*
Light: *100 Suns*
Mack: *Passport to the Cosmos*
Roberts: *Science Fiction*
Schlemmer: *The Only Planet of Choice*
Sheckley: *The People Trap*
Spufford: *Backroom Boys*
Stapledon: *Star Maker*
Vallée: *Anatomy of a Phenomenon*
Zimmerman: *Encountering Alien Otherness*

Interview with Mike Jay

Chapter 3: Betty and Barney Hill

Bennett: *Looking for Orthon*
Bowen: *The Humanoids*
Evans and Stacy: *The UFO Mystery*
Fuller: *The Interrupted Journey*
Jacobs: *UFOs and Abductions*
Jung: *Flying Saucers*
Leslie and Adamski: *Flying Saucers Have Landed*

Interview with Budd Hopkins

Chapter 4: Alien Surgery

Bear: *Blood Music*
Davis: *TechGnosis*

Evans and Stacy: *The UFO Mystery*
Good: *Unearthly Disclosure*
Hansen: *Missing Times*
Jacobs: *The Threat*
Jacobs: *UFOs and Abductions*
Jong: *Fear of Flying*
LeGuin: *The Left Hand of Darkness*
Leir: *The Aliens and the Scalpel*
Strieber: *Communion*
Vallée: *Anatomy of a Phenomenon*
Watson: *Miracle Visitors*
Wilson: *Alien Dawn*

Interview with Budd Hopkins
Interview with Chris French

Chapter 5: John Mack
Mack: *Abduction*
Mack: *Passport to the Cosmos*

Interview with John Mack

Chapter 6: Alien Cover-Up
Bruni: *You Can't Tell the People*
Couper and Henbest: *Mars*
Darlington: *The Dreamland Chronicles*
Evans and Stacy: *The UFO Mystery*
Hansen: *Missing Times*
Hynek: *The Hynek UFO Report*
Hynek: *The UFO Experience*
Jacobs: *UFOs and Abductions*
Klass: *UFOs, The Public Deceived*
Pope: *Open Skies*
Saler, Ziegler and Moore: *UFO Crash at Roswell*
Strugatsky and Strugatsky: *Roadside Picnic*
Vallée: *Anatomy of a Phenomenon*

Zimmerman: *Encountering Alien Otherness*

Interview with Graham Birdsall
Interview with Georgina Bruni
Interview with Nick Pope

Chapter 7: Marshall Applewhite
Clarke: *Rendezvous with Rama*
Davis: *TechGnosis*
Jacobs: *UFOs and Abductions*
Saler, Ziegler and Moore: *UFO Crash at Roswell*

Chapter 8: Aliens and Angels
Adams: *The Hitchhiker's Guide to the Galaxy*
Angelucci: *Secrets of the Saucers*
Blish: *A Case of Conscience*
Brin: *Sundiver*
Coyne: *The 'Many Worlds' and Religion*
Harpur: *The Philosophers' Secret Fire*
Jacobs: *UFOs and Abductions*
Jay: *The Air Loom Gang*
Jung: *Flying Saucers*
Lem: *Fiasco*
Marciniak: *Bringers of the Dawn*
Schlemmer: *The Only Planet of Choice*
Sheckley: *The People Trap*
Thompson: *Angels and Aliens*
Turkle: *Life on the Screen*
Von Däniken: *Chariots of the Gods*
Von Däniken: *Return to the Stars*
Waugh: *The Ordeal of Gilbert Pinfold*
Wilson: *Alien Dawn*
Wordsworth: *Poetical Works*

Chapter 9: Philip Kindred Dick
Davis: *TechGnosis*
Dick: *Our Friends from Frolix 8*
Dick: *The Three Stigmata of Palmer Eldritch*
Roberts: *Science Fiction*

Erik Davis, interview with Dick at
http://frontwheeldrive.com/philip_k_dick.html

Chapter 10: Alien Suggestion
Jacobs: *UFOs and Abductions*

Interview with David Oakley
Interview with Chris French

Chapter 11: Claudius Ptolemaeus/Galileo Galilei
Bear: *Blood Music*
Monod: *Chance and Necessity*
Shelley: *Frankenstein*

Chapter 12: Alien Eyes
Adams: *The Hitchhiker's Guide to the Galaxy*
Aldiss: *The Penguin Science Fiction Omnibus*
Bowen: *The Humanoids*
Bruni: *You Can't Tell the People*
Cadigan: *Mindplayers*
Crowe: *The Extraterrestrial Life Debate*
Dick, P.: *The Three Stigmata of Palmer Eldritch*
Dick, S.: *Plurality of Worlds*
Fort: *The Book of the Damned*
Gernsback: *Ralph 124C 41+*
Gibson: *Neuromancer*
Harpur: *The Philosophers' Secret Fire*
Heard: *Is Another World Watching?*
Jacobs: *The Threat*

Chapter 15: Marvin Minsky/Steve Jobs
Turkle: *Life on the Screen*
Vonnegut: *The Sirens of Titan*

Chapter 16: Alien Metal
Aldiss and Wingrove: *Trillion Year Spree*
Amis and Conquest: *Spectrum*
Ashbery: *Selected Poems*
Bowen: *The Humanoids*
Clarke: *Rendezvous with Rama*
Featherstone and Burrows: *Cyberspace/Cyberbodies/Cyberpunk*
Gibson: *Neuromancer*
Gray: *Cyborg Citizen*
Grenville: *The Uncanny*
Mazlish: *The Fourth Discontinuity*
Roberts: *Science Fiction*
Turkle: *Life on the Screen*

Chapter 17: Stanislaw Lem
Aldiss and Wingrove: *Trillion Year Spree*
Lem: *Fiasco*
Lem: *Imaginary Magnitudes*
Lem: *Solaris*
Mann: *Mammoth Encyclopaedia*

Conclusion: The Snow Man
Adams: *The Hitchhiker's Guide to the Galaxy*
Amis: *New Maps of Hell*
Bloom: *Omens of Millennium*
Bradbury: *The Martian Chronicles*
Crowe: *The Extraterrestrial Life Debate*
Dick, P.: *The Three Stigmata of Palmer Eldritch*
Dick, S.: *Plurality of Worlds*
Grenville: *The Uncanny*
Harpur: *The Philosophers' Secret Fire*

Harrison: *After Contact*
Kafka: *Metamorphosis*
Lamb: *The Search for Extraterrestrial Intelligence*
Leslie and Adamski: *Flying Saucers Have Landed*
Lindsay: *A Voyage to Arcturus*
Mack: *Passport to the Cosmos*
Monod: *Chance and Necessity*
Roberts: *Science Fiction*
Sheckley: *The People Trap*
Stevens: *Collected Poetry and Prose*
Stevenson: *Dr Jekyll and Mr Hyde*
Strugatsky and Strugatsky: *Roadside Picnic*
Vonnegut: *The Sirens of Titan*
Wells: *The Invisible Man*
Wordsworth: *Poetical Works*

Sources

On the subject of aliens, no list of sources can hope to be complete or even uncontentious. Almost as soon as I began to seek out books on the subject, I was overwhelmed by the sheer quantity of what was available. What follows, therefore, is my entirely subjective selection. First, there is a list of books, films and television shows that have directly contributed to this book. I have given publication dates of the editions I used; in the text, where necessary, I have provided the original publication dates. Secondly, I have allocated textual sources to individual chapters. Finally, I have provided a list of web sites. The web being what it is, this last is a perilous undertaking. For every web site I list there will be thousands of others, equally helpful or equally misleading. Furthermore, sites change radically or disappear all the time. The ones I list are simply those that worked for me at the time.

Books

Adams, Douglas, *The Hitchhiker's Guide to the Galaxy: The Trilogy of Four*, Picador, London 2002

Aldiss, Brian, ed., *The Penguin Science Fiction Omnibus*, Penguin, London, 1973

Aldiss, Brian and Wingrove, David, *Trillion Year Spree: The History of Science Fiction*, Paladin, London, 1988

Amis, Kingsley, *New Maps of Hell*, New English Library, London 1969

Amis, Kingsley and Conquest, Robert, eds., *Spectrum: A Science Fiction Anthology*, Victor Gollancz, London, 1961

Angelucci, Orfeo, *The Secret of the Saucers,* Amherst Press, Amherst, WI, 1955

Ashbery, John, *Selected Poems*, Carcanet, Manchester, 1998

Asimov, Isaac, *The Gods Themselves*, Gollancz, London, 2000

Bear, Greg, *Blood Music*, Gollancz, London, 2002

Bennett, Colin, *Looking for Orthon: The Story of George Adamski, the First Flying Saucer Contactee and How He Changed the World*, Paraview Press, London, 2001

Blish, James, *A Case of Conscience*, Gollancz, London, 1999

Bloom, Harold, *Omens of Millennium*, Fourth Estate, London 1996

Bova, Ben and Preiss, Byron, eds, *Are We Alone in the Cosmos? The Search for Alien Contact in the New Millennium*, ibooks, New York, 1999

Bowen, Charles, ed., *The Humanoids*, Futura, London 1977

Bradbury, Ray, *The Martian Chronicles*, Granada, London, 1983

Bradbury, Ray, *The Vintage Bradbury*, introduction by Gilbert Highet, Vintage Books, New York, 1990

Brin, David, *Sundiver*, Orbit, London, 1996

Bruni, Georgina, *You Can't Tell the People: The Cover-Up of Britain's Roswell*, Pan Books, London, 2001

Cadigan, Pat, *Mindplayers*, Gollancz, London, 2000

Cadigan, Pat, ed., *The Ultimate Cyberpunk*, ibooks, New York, 2002

Clarke, Arthur C., *Rendezvous with Rama*, Bantam, London, 1990

Couper, Heather and Henbest, Nigel, *Mars: The Inside Story of the Red Planet*, Headline, London, 2001

Coyne, George V. SJ, *The 'Many Worlds' and Religion*, Vatican Observatory, Tucson, 1997

Crowe, Michael J., *The Extraterrestrial Life Debate 1750–1900:*

The Idea of a Plurality of Worlds from Kant to Lowell, Cambridge University Press, Cambridge, 1986

Darlington, David, *The Dreamland Chronicles, The Legends of Area 51 – America's Most Secret Military Base*, Warner Books, London, 1999

Davies, Paul, *Are We Alone? Implications of the Discovery of Extraterrestrial Life*, Penguin, London, 1995

Davis, Erik, *TechGnosis, Myth, Magic and Mysticism in the Age of Information*, Serpent's Tail, London, 1999

Dick, Philip K., *Our Friends from Frolix 8*, Voyager, London, 1997

Dick, Philip K., *The Three Stigmata of Palmer Eldritch*, Gollancz, London, 2003

Dick, Steven J., *Life on Other Worlds: The Twentieth Century Extraterrestrial Life Debate*, Cambridge University Press, Cambridge, 1998

Dick, Steven J., *Plurality of Worlds: The Origins of the Extraterrestrial Life Debate from Democritus to Kant*, Cambridge University Press, Cambridge, 1982

Evans, Hilary and Stacy, Dennis, eds, *The UFO Mystery, The 50-Year Quest to Solve the World's Greatest Enigma*, John Brown, London, 1998

Featherstone, Mike and Burrows, Roger, eds, *Cyberspace/Cyberbodies/Cyberpunk: Culture of Technological Embodiment*, Sage, London, 1995

Fort, Charles, *The Book of the Damned*, revised by X, introduction by Bob Rickard, John Brown, London, 1995

Fuller, John G., *The Interrupted Journey*, Dell, New York, 1987

Gernsback, Hugo, *Ralph 124C 41+: A Romance of the Year 2660*, introduction by Jack Williamson, University of Nebraska Press, Lincoln, 2000

Gibson, William, *Neuromancer*, Voyager, London, 1995

Good, Timothy, *Unearthly Disclosure: Conflicting Interests in the Control of Extraterrestrial Intelligence*, Century, London, 2000

Gray, Chris Hables, *Cyborg Citizen: Politics in the Posthuman Age*, Routledge, London, 2002

Grenville, Bruce, ed., *The Uncanny: Experiments in Cyborg*

Culture, Vancouver Art Gallery/Arsenal Pulp Press, Vancouver, 2002

Hansen, Terry, *The Missing Times: News Media Complicity in the UFO Cover-Up*, www.Xlibris.com, 2000

Harpur, Patrick, *The Philosophers' Secret Fire: A History of the Imagination*, Penguin, London, 2002

Harrison, Albert, *After Contact: The Human Response to Extraterrestrial Life*, Perseus, Cambridge, 2002

Heard, Gerald, *Is Another World Watching? The Riddle of the Flying Saucers*, Harper & Bros., New York, 1951

Holroyd, Stuart, *Prelude to the Landing on Planet Earth*, W. H. Allen, London, 1977

Hopkins, Budd, *Witnessed: The True Story of the Brooklyn Bridge Abductions*, Bloomsbury, London, 1997

Howard, L. G., *If in Doubt Blame the Aliens! A New Scientific Analysis of UFO Sightings, Alleged Alien Abductions, Animal Mutilations and Crop Circles*, Writer's Showcase, Lincoln, NE, 2000

Hoyle, Fred and Wickramasinghe, Chandra, *Lifecloud: The Origin of Life in the Universe*, J. M. Dent & Sons, London, 1978

Hynek, J. Allen, *The UFO Experience: A Scientific Enquiry*, McGraw-Hill, New York, 1972

Hynek, J. Allen, *The Hynek UFO Report*, Souvenir Press, London, 1998

Jacobs, David M., *The Threat: Revealing the Secret Alien Agenda*, Simon and Schuster, New York, 1999

Jacobs, David M., *UFOs and Abductions: Challenging the Borders of Knowledge*, University Press of Kansas, Lawrence, 2000

Jay, Mike, *The Air Loom Gang: The Strange and True Story of James Tilly Matthews and His Visionary Madness*, Bantam Press, London, 2003

Jong, Erica, *Fear of Flying*, HarperCollins, London, 1998

Jung, Carl Gustav, *Flying Saucers: A Modern Myth of Things Seen in the Sky*, Routledge, London, 2002

Kafka, Franz, *Metamorphosis and Other Stories*, Penguin, London, 1975

Klass, Philip J., *UFOs: The Public Deceived*, Prometheus Books, New York, 1993

King, Stephen, *Danse Macabre*, Warner Books, London, 1993

Lamb, David, *The Search for Extraterrestrial Intelligence: A Philosophical Inquiry*, Routledge, London, 2001

LeGuin, Ursula K., *The Left Hand of Darkness*, Panther, London, 1973

Leir, Roger K., *The Aliens and the Scalpel: Scientific Proof of Extraterrestrial Implants in Humans*, Granite Publishing, Columbus, 1998

Lem, Stanislaw, *Fiasco*, trans. Michael Kandel, Harcourt Brace Jovanovich, New York, 1988

Lem, Stanislaw, *Imaginary Magnitudes*, trans. Marc E. Heine, Harcourt Brace Jovanovich, New York, 1984

Lem, Stanislaw, *Solaris*, Faber & Faber, London, 2003

Leslie, Desmond and Adamski, George, *Flying Saucers Have Landed*, Werner Laurie, London, 1953

Lewis, Tony, ed., *The Best of Astounding*, Baronet, New York, 1978

Light, Michael, *100 Suns, 1945–1962*, Jonathan Cape, London, 2003

Lindsay, David, *A Voyage to Arcturus*, introduction by J. B. Pick, Canongate, Edinburgh, 1998

Mack, John E., *Abduction: Human Encounters with Aliens*, Charles Scribner's Sons, New York, 1994

Mack, John E., *Passport to the Cosmos: Human Transformation and Alien Encounters*, Thorsons, London, 2000

Mann, George, ed., *The Mammoth Encyclopaedia of Science Fiction*, Robinson, London, 2001

Marciniak, Barbara, *Bringers of the Dawn: Teachings from the Pleiadians*, Bear & Co., Rochester, NY, 1992

Mazlish, Bruce, *The Fourth Discontinuity: The Co-Evolution of Humans and Machines*, Yale University Press, New Haven, 1993

Monod, Jacques, *Chance and Necessity: An Essay on the Natural Philosophy of Modern Biology*, trans. Austryn Wainhouse, Fontana, London, 1974

Poe, Edgar Allan, *The Narrative of Arthur Gordon Pym of Nantucket*, Penguin, London, 1975

Pope, Nick, *Open Skies, Closed Minds: For the First Time a Government UFO Expert Speaks Out*, Dell, New York, 2000

Roberts, Adam, *Science Fiction*, Routledge, London, 2000

Saler, Benson, Ziegler, Charles A. and Moore, Charles B., *UFO Crash at Roswell: The Genesis of a Modern Myth*, Smithsonian Institution, Washington, DC, 1997

Schlemmer, Phyllis V., transceiver, *The Only Planet of Choice: Essential Briefings from Deep Space*, ed. Mary Bennett, Gateway, Dublin, 1996

Sheckley, Robert, *Mindswap*, Ace Books, New York, 1978

Sheckley, Robert, *The People Trap*, Pan Books, London, 1972

Shelley, Mary, *Frankenstein*, Penguin, London, 1994

Shiel, M. P., *The Purple Cloud*, introduction by John Clute, University of Nebraska Press, Lincoln, 2000

Shippey, Tom, ed., *The Oxford Book of Science Fiction Stories*, Oxford University Press, Oxford, 1992

Sladek, John, *The Muller-Fokker Effect*, Panther, London, 1972

Spufford, Francis, *Backroom Boys: The Secret Return of the British Boffin*, Faber & Faber, London, 2003

Stapledon, Olaf, *Star Maker*, Gollancz, London, 2003

Stevens, Wallace, *Collected Poetry and Prose*, Library of America, Washington, DC, 1997

Stevenson, Robert Louis, *Dr Jekyll and Mr Hyde and Other Stories*, Penguin, London, 1979

Strieber, Whitley, *Communion: A True Story, Encounters with the Unknown Organization*, Arrow, London, 1988

Strugatsky, Arkady and Strugatsky, Boris, *Roadside Picnic*, Gollancz, London, 2000

Thompson, Keith, *Angels and Aliens: UFOs and the Mythic Imagination*, Fawcett Columbine Ballantine Books, New York, 1993

Turkle, Sherry, *Life on the Screen: Identity in the Age of the Internet*, Touchstone, New York, 1997

Vallée, Jacques, *Anatomy of a Phenomenon: UFOs in Space*, Tandem, London, 1974

Von Däniken, Erich, *Chariots of the Gods: Was God an*

Astronaut?, Souvenir Press, London, 1989

Von Däniken, Erich, *Return to the Stars: Evidence for the Impossible*, trans. Michael Heron, Souvenir Press, London, 1970

Vonnegut, Kurt, *The Sirens of Titan*, Gollancz, London, 2003

Watson, Ian, *Miracle Visitors*, Gollancz, London, 2003

Waugh, Evelyn, *The Ordeal of Gilbert Pinfold and Other Stories*, Chapman & Hall, London, 1975

Wells, H. G., *The Invisible Man*, Signet, New York, 2002

Wells, H. G., *The War of the Worlds*, introduction by Arthur C. Clarke, The Modern Library, New York, 2002

Wilson, Colin, *Alien Dawn: An Investigation into the Contact Experience*, Virgin, London, 1998

Wordsworth, William, *The Poetical Works of William Wordsworth*, Oxford University Press, Oxford, 1926

Zimmerman, Michael, *Encountering Alien Otherness in The Concept of the Foreign*, Lexington Books, Lanham, 2003

Film and Television

A.I., Steven Spielberg, 2001

Alien, Ridley Scott, 1979

Alien Resurrection, Jean-Pierre Jeunet, 1997

The Andromeda Strain, Robert Wise, 1971

Apocalypse Now, Francis Ford Coppola, 1979

Blade Runner, Ridley Scott, 1982

Close Encounters of the Third Kind, Steven Spielberg, 1977

Contact, Robert Zemeckis, 1997

The Day the Earth Stood Still, Robert Wise 1951

Dune, David Lynch, 1984

Earth Versus the Flying Saucers, Fred F. Sears, 1956

E.T. the Extraterrestrial, Steven Spielberg, 1982

Explorers, Joe Dante, 1985

Forbidden Planet, Fred McLeod Wilcox, 1956

Galaxy Quest, Dean Parisot, 1999

Gremlins, Joe Dante, 1984

Independence Day, Roland Emmerich, 1996

Invaders from Mars, William Cameron Menzies, 1953

Invasion of the Body Snatchers, Don Siegel, 1955
It Came from Beneath the Sea, Robert Gordon, 1955
Killers from Space, W. Lee Wilder, 1954
K-Pax, Iain Softley, 2001
Tha Manchurian Candidate, John Frankenheimer, 1962
The Matrix, Andy and Larry Wachowski, 1999
Men in Black, Barry Sonnenfeld, 1997
Metropolis, Fritz Lang, 1926
The Quatermass Experiment, Val Guest, 1955
Quatermass 2, Val Guest, 1957
Red Planet, Antony Hoffman, 2000
The Right Stuff, Philip Kaufman, 1983
Rosemary's Baby, Roman Polanski, 1968
Saturn 3, Stanley Donen, 1980
Solaris, Andrei Tarkovsky, 1972
Solaris, Steven Soderbergh, 2002
Star Trek, Gene Roddenberry, 1966 –
Star Trek: First Contact, Jonathan Frakes, 1996
Star Trek Generations, David Carson, 1994
Star Trek: Nemesis, Stuart Baird, 2002
Star Wars, George Lucas, 1977
The Stepford Wives, Bryan Forbes, 1974
Taken, Breck Eisner and others, 2002
The Terminator, James Cameron, 1984
Terminator 2: Judgment Day, James Cameron, 1991
Terminator 3: The Rise of the Machines, Jonathan Mostow,
   2003
Them!, Gordon Douglas, 1954
The Thing from Another World, Christian Nyby, 1951
This Island Earth, Joseph Newman, 1955
Twilight Zone: the Movie, John Landis, Steven Spielberg, Joe
   Dante, George Miller, 1983
2001: A Space Odyssey, Stanley Kubrick, 1968
2010, Peter Hyams, 1984
The UFO Incident, Richard A. Colla, 1975
Westworld, Michael Crichton, 1973
The X-Files, Chris Carter, 1993–2002

Selected Web sites

aj-net.org
clavius.as.arizona.edu
www.boblazar.com
www.daniken.com
www.etcontact.net
www.gafintl-adamski.com
www.history-mystery.com
www.intrudersfoundation.org
www.magicdragon.com
www.martiansgohome.com
www.mufon.com
www.nasa.gov
www.newadvent.org
www.nickpope.net
www.nidsci.org/
www.prufos.co.uk
www.rickross.com
www.seti-inst.edu
www.space.com
www.susanblackmore.co.uk/
www.ufology.org.uk
www.unarius.org
www.wave.net
www.worldofthestrange.com

Index